WITHD

D1631507

MONOGRAPHS ON SCIENCE, TECHNOLOGY, AND SOCIETY

MONOGRAPHS ON SCIENCE, TECHNOLOGY, AND SOCIETY

1 Eric Ashby and Mary Anderson: The politics of clean air
2 Edward Pochin: Nuclear radiation: risks and benefits
3 L. Rotherham: Research and innovation: a record of the Wolfson
    Technological Projects Scheme 1968–1981; with a foreword and
    postscript by Lord Zuckerman
4 John Sheail: Pesticides and nature conservation: the British
    experience 1950–1975

PESTICIDES AND NATURE CONSERVATION

The British Experience 1950–1975

JOHN SHEAIL

*Principal Scientific Officer, Institute
of Terrestrial Ecology, Monks Wood
Experimental Station, Huntingdon*

CLARENDON PRESS · OXFORD
1985

Oxford University Press, Walton Street, Oxford OX2 6DP
Oxford New York Toronto
Delhi Bombay Calcutta Madras Karachi
Kuala Lumpur Singapore Hong Kong Tokyo
Nairobi Dar es Salaam Cape Town
Melbourne Auckland
and associated companies in
Beirut Berlin Ibadan Mexico City Nicosia

OXFORD is a trade mark of Oxford University Press

Published in the United States
by Oxford University Press, New York

British Library Cataloguing in Publication Data
Sheail, John
Pesticides and nature conservation: the British
experience, 1950–1975.—(Monographs on science,
technology and society)
1. Pesticides and wildlife—Great Britain
I. Title   II. Series
639.9'0941   QL82
ISBN 0-19-854150-3

Library of Congress Cataloging in Publication Data
Sheail, John.
Pesticides and Nature Conservation.
(Monographs on science, technology, and society)
Bibliography: p.
Includes index.
1. Pesticides—Environmental aspects—Great
Britain. I. Title. II. Series.
QH545.P4S5 1985   363.7'384   85-8845
ISBN 0-19-854150-3

Set by Cotswold Typesetting, Cheltenham
Printed in Great Britain by
St. Edmundsbury Press,
Bury St. Edmunds, Suffolk

PREFACE

With the publication of the report on *Agriculture and pollution* by the Royal Commission on Environmental Pollution in 1979, the time seemed ripe for an historical review of the circumstances in which the threat of pesticides to wildlife populations was first recognized, and how administrators and scientists responded to that perceived threat given the resources and constraints of the period. The Institute of Terrestrial Ecology, which is part of the Natural Environment Research Council, was well placed to carry out such an historical reconstruction. The Institute had absorbed the research stations and staff of the former Nature Conservancy in 1973, including the Monks Wood Experimental Station where most of the Conservancy's pioneer pesticide/wildlife research was done. The Director of the Institute, Mr J.N.R. Jeffers, suggested that a book should be written, based not only on published material and the recollections of those who played a key role in the events described, but also on the files and papers that survived at Monks Wood and in the registry of the former Nature Conservancy.

The outcome is the present book. The author is an historical geographer, who has been a member of the Monks Wood Experimental Station for many years. He has not been directly involved in the research on pesticides and wildlife.

The book illustrates how a chemical revolution took place on Britain's farms during the 1950s and 1960s. Most of the damage inflicted on wildlife was caused by a small number of relatively persistent compounds used in insecticides. Research programmes were set up to investigate how different forms of wildlife were being affected. The voluntary controls already imposed on the use of pesticides were continually reappraised against a background of mounting public concern. The experience gained in this field of applied ecology was to have important repercussions for the regulation of environmental pollution generally.

Because of the range of material available to the author, this book is necessarily written from a conservation viewpoint. Many years will elapse before the full documentation of the pesticide and farming industries becomes available. In the meantime, this book should provide some insight into how research and executive action over the effects of pesticides on wildlife populations came about.

It is a pleasure to record all the help given me by those whose work is described in the book, and who have confirmed, corrected, and extended what I might have deduced from the written record. I am grateful to Sir Hermann Bondi, Dr J.C. Bowman, and Lord Ashby, who provided much-needed encouragement during the book's preparation. Gillian Sheail, Dr D. Osborn, and Mr T.C.E. Wells commented on drafts of the text. I am grateful to *Punch* and to the syndics of Cambridge University Library for permission to reproduce Fig. 12.1. Mr. J. Williamson prepared all the other figures. The facilities extended to me by the Nature Conservancy Council for looking at the registry of the former Nature Conservancy are gratefully acknowledged.

Monks Wood   J.S.
December 1984

CONTENTS

A CHRONOLOGY OF SELECT EVENTS

1949   Establishment of the Nature Conservancy. National Parks and Access to the Countryside Act.

1951   Report of the first Zuckerman Working Party.

1953   Report of the second Zuckerman Working Party. Appointment of the Inter-departmental Advisory Committee on Poisonous Substances.

1955   Ministry of Transport Circular 718 published. Report of the third Zuckerman Working Party. Extension of the terms of reference and membership of the Advisory Committee.

1956   First reported deaths linked with seed-dressings.

1957   Introduction of voluntary Notification Scheme (later Pesticides Safety Precautions Scheme).

1959

January   Memorandum by Director-General of Nature Conservancy on setting up an applied research and demonstration station. Decision to establish a Toxic Chemicals Unit in Conservancy.

July   Huntingdon Working Party formed.

November Appointment of (Sanders) Research Study Group.

December First meeting of Wildlife Panel of Advisory Committee.

1960

January   Toxic Chemicals and Wild Life Section established in Conservancy.

May   Treasury approval in principle for new station.

August   Joint Committee of BTO and RSPB formed to investigate deaths caused by seed-dressings. Conservancy report on the incidence and causes of fox-death.

November Conservancy's new station named the Monks Wood Experimental Station.

1961

Spring   Peak of bird mortality caused by seed-dressings. First BTO survey of peregrine falcon.

April   First Shackleton Debate.

June   Report of Commons Select Committee on Estimates on the
                  Ministry of Agriculture.
                Voluntary ban imposed on use of aldrin, dieldrin, and
                  heptachlor in spring seed-dressings.
September Report of (Sanders) Research Study Group.

1962
March   Contract for building of Monks Wood agreed.
June–July   'Silent spring' published in *New Yorker.*

1963
January   Treasury approval for 12 additional posts in Toxic
                    Chemicals and Wild Life Section.
February   Publication of British edition of 'Silent spring'.
                  Second Shackleton Debate.
May   First National Nature Week.
                  Publication of 'Chemicals for the gardener' by Ministry of
                    Agriculture.
                  Announcement of a further examination of pesticide use by
                    Advisory Committee.
June   Report of the President's Science Advisory Committee.
October   Official opening of Monks Wood.

1963
November First 'Countryside in 1970' conference.

1963/64
winter   Merthyr Tydfil and Smarden incidents involving
                    fluoroacetamide.

1964
March   First report of (Frazer) Research Committee.
                  Report of Advisory Committee on persistent
                    organochlorine pesticides.
                  Announcement of further restrictions on the use of aldrin,
                    dieldrin, and heptachlor.

1965
March   Royal Assent given to Science and Technology Bill, and
                    setting up of the Natural Environment Research Council.
July   NATO-sponsored Advanced Study Institute on Pesticides
                    held at Monks Wood.
September Wisconsin meeting on peregrine falcon.
November Second 'Countryside in 1970' conference.

1966   Publication of first evidence of presence of PCBs in the
              environment.

1967
  January   Publication of Advisory Committee's 'Review of the
              present safety arrangement for the use of toxic chemicals
              in agriculture and food storage'.
  March   Publication of Kenneth Mellanby's 'Pesticides and
              pollution'. *Torrey Canyon* disaster.

1969
  Sept/Nov. Irish Sea seabird wreck.
  December   Publication of Advisory Committee's 'Further review of
              certain persistent organochlorine pesticides'.

1970   *European Conservation Year*
  January   Visit of Prime Minister to Monks Wood.
  May   White Paper on pollution.
  October   Third 'Countryside in 1970' conference.
  November   Visit of HRH Prince of Wales to Monks Wood.

1971
  February   First report of the Royal Commission on Environmental
              Pollution.
  November   Green Paper on a 'Framework of government research and
              development', and publication of Dainton and Rothschild
              reports.

1972
  April   Announcement that no legislation on pesticide use
              contemplated.
  July   White Paper on 'The framework of government research
              and development', and announcement of the Split.

1973
  August   Withdrawal announced of all seed-dressings containing
              aldrin and dieldrin.
  November   Nature Conservancy Council and Institute of Terrestrial
              Ecology come into being.

1979
  September Report of the Royal Commission on Environmental
              Pollution on 'Agriculture and pollution'.

1

INTRODUCTION

Ecologists have always emphasized the interdependence of living organisms. Each species forms part of a food chain or cycle; each affects the living space of another. None leads a completely independent existence. In that sense, the death of a plant or animal is no isolated incident—it is bound to have wider ramifications. The term 'species network' has been coined to describe the interlocking system (Elton and Miller 1954). It is a network that embraces human activity. Even the time-honoured custom of collecting dead wood for fuel could have a significant effect on the dispersal and interchange of species (Elton 1966), and it would, moreover, be misleading to portray it as an isolated incident in time and space—it could lead to further, inevitable changes in the network.

The scale of human influence on the natural environment has increased dramatically over the last few decades and has called for a deeper ecological understanding of the interlocked system. Attempts have been made to apply the lessons of the past to the future. As well as knowing how, when, and where changes in land use and resource management have occurred, it is necessary to know *why* these changes have been promoted or tolerated. The ecological reckoning has had to take account of the interplay of political, administrative, and the wider, scientific perspectives and influence. The role of the different interest and pressure groups in the decision-making process has to be heeded. In short, it is necessary to explore what some have called 'the web of human ecology'.

This book describes how the effects of pesticides on the interlocking system were perceived by the nature conservation movement in Britain, and how a scientific, administrative, and political response was made. It is a story of how one, perhaps casual, observation led to another, and how information was brought together and exploited. A rudimentary response became increasingly complex and wide-ranging. The story has no beginning and no end. Pesticide users and nature conservationists approached the issues with the preconceptions of the past, and many of the problems remain unresolved. The story also has no limits; this account is simply a British version of a dilemma that is being confronted all over the world. It provides a case study of man's competence in managing the world in which he and over two million other species subsist.

So far as the British story of pesticides and the nature conservation movement is concerned, 1970 is likely to remain an important point of reference. It was European Conservation Year, and preparations were under way for the United Nations Conference on the Environment at Stockholm in 1972. The British Government published a White Paper in 1970 on *The protection of the environment: the fight against pollution.* It was the first to appear on the subject. A few months earlier, a Royal Commission on Environmental Pollution had been appointed. In these and other ways, the Government sought to demonstrate how Britain was active both at the international level and in preparing 'its own environmental defences'.

According to the White Paper, pollution affected the physical health of the population and 'the ordinary pleasure and contentment of people in the quality of their life'. It was not necessary to wait until they became ill before taking action. In order to improve and maintain public health, the Government had to anticipate health hazards and respond to those that had occurred. There were four prerequisites, namely better scientific and technological knowledge, economic priorities and economic decisions, the correct legal and administrative framework, and the will to do the job. It was for the Government to take the lead, but success also depended 'on an increasingly informed and active public opinion' (Secretary of State 1970).

The authors of the White Paper considered that more research was required on pollutants themselves, on their effects on human health and amenity, and on their general influence on plant and animal communities. The White Paper emphasized that, 'because ecological systems are complicated and change continually in response to many factors other than pollution, we need a background of basic and often long-term ecological knowledge before we can accurately interpret all the actions of pollutants'. Both the White Paper and the reports of the Royal Commission on Environmental Pollution emphasized that a great deal was being done in this direction.

The large scale use of chemical pesticides provided one of the first indications that a chemical revolution was taking place in the British countryside during the 1950s and 1960s. Unfortunately, the scope for writing a detailed account of how this revolution took place, and of how pesticides came to be used and regulated, is limited by the lack of primary information. The relevant files and papers of the pesticide and farming industries are likely to remain confidential for many years to come. This makes it much harder to discover how decisions were actually taken on the formulation, deployment, and control of these pesticides.

This book focuses on one aspect of pesticide use, namely, the impact of pesticides on wildlife and the natural environment. Not only was this an increasingly important aspect of the chemical revolution in the countryside, but the perception of the effects of pesticides, and the responses made by the administrator and scientist, may offer insights into what was happening more generally in the field of pesticide use and regulation. At the very least, this book should provide an interim statement until the primary documentation in official and private archives becomes more generally available.

The dangers and inconveniencies arising from pollution touched on many interests. A wide range of departments and agencies, in both central and local government, was involved in taking preventative and remedial action. One of these agencies was the Nature Conservancy. Together with the voluntary nature conservation bodies, the Conservancy attached considerable importance to the protection of wildlife and the natural environment from the adverse effects of pollution. It is the purpose of this book to identify the role of the Nature Conservancy and the wider conservation movement in highlighting the effects of pesticides and pollution on wildlife.

The book is largely based on the papers that survive in the Conservancy's registry and the files of individual scientists. It describes how the effects of pesticides were first perceived, and the response made within the opportunities and constraints of the period. Although some aspects of the story must remain confidential, particularly where these relate to other organizations, it is hoped that the study will contribute to a wider and deeper understanding of how research and executive action on the effects of pesticides and pollution on wildlife developed.

Although the historian must reconstruct the events of the past as coherently as possible, he has to avoid giving a false sense of coherence and tidiness to those events. This is by no means easy. Furthermore, documentary sources do not provide a complete and objective record of what happened. When Charles Elton once saw some minutes of a meeting he had attended, he commented that they gave him the 'impression of a committee of high-powered men of action letting nothing stand in their way'. Alas, this was quite a misleading impression. In his words, the minutes were 'a true record of what we should have liked the meeting to have discussed and decided'. Elton's remarks underline the need to search out and compare as many different kinds of historical information as possible. For this and other reasons, the recollections of the administrators and scientists involved in the events described can provide an invaluable means of correcting and supplementing the impressions gained from documentary sources and their interpretation.

In concentrating on the activities and personnel of the Nature Conservancy, it is not intended to provide a history of that organization. The pesticides issue became, however, so important to the Conservancy that the aims and methods used to influence the scale and character of pesticide use provide insights into the wider objectives and character of the

Conservancy over the 24 years of its existence. The Nature Conservancy was created under a Royal Charter in March 1949, with the following terms of reference:

to provide scientific advice on the conservation and control of the natural flora and fauna of Great Britain; to establish, maintain and manage nature reserves in Great Britain, including the maintenance of physical features of scientific interest; and to organize and develop the scientific services related thereunto.

The Conservancy derived its statutory powers from the National Parks and Access to the Countryside Act of the same year (Sheail 1976, 1984).

The Conservancy became the fourth research council, joining the Department of Scientific and Industrial Research (DSIR), created in 1919, the Medical Research Council, and the Agricultural Research Council, both of which were granted Royal Charters in 1920 and 1931 respectively. Whilst they were formally subject to ministerial directives and were mainly dependent on their grant-in-aid from the Treasury, the Councils enjoyed a considerable degree of autonomy in selecting and implementing their research programmes. This freedom was embodied in what was known as 'the Haldane principle'. For the most part, the governing body of each Council consisted of independent members drawn from the scientific and other fields (Pile 1979). The term 'council' was not used in the Conservancy; the governing body was usually called 'the Nature Conservancy Committee'. The 'England Committee' and the 'Scientific Policy Committee' were two of the advisory committees. The most senior officer was the Director-General.

In its Annual Report of 1953, the Conservancy conceded that its task was 'immensely complex both scientifically and administratively'. The initial and preparatory stages had taken much longer than anticipated. Because the creation of such a body as the Nature Conservancy had been long overdue, there were 'heavy arrears of site survey and preparation and of training and investigation'. Fresh demands and commitments were arising all the time. The conservation and research programmes had to be adapted to take account of such unexpected events as the North Sea Floods of early 1953 and the rapid spread of myxomatosis among the rabbit population soon afterwards.

The Conservancy's capacity to respond to its Charter functions was further inhibited by the economies imposed on public expenditure. As early as 1954, the Annual Report complained of how 'severe economy will have to be exercised in respect of all the Conservancy's research and other activities to keep within the financial provision'. The Annual Report of 1955 warned that 'failure to provide the necessary funds for research and conservation including the acquisition of nature reserves could only result in loss or damage which would be felt by science and by the community many years after the relatively minute contribution made by any consequent savings in the national budget had been forgotten'.

Such warnings highlighted the Conservancy's weak position. Only the informed minority saw 'the potential value of the scientific, educational and Reserve Management activities'. The wider public would remain ignorant and unimpressed until the Conservancy had achieved something, and had publicized it with 'simple and effective publications and illustrations'. Paradoxically, this could only come about with more funds and greater manpower.

In tracing the way in which the threat of pesicides to wildlife was perceived, and the variety of responses made to that threat, close reference must be made to the more general preoccupations and constraints of the nature conservation movement, and of the Nature Conservancy in particular. It was never possible to tackle the dangers posed by toxic chemicals in isolation. Nature conservation was only one of a large number of considerations that had to be taken into account in the management of the countryside. The importance attached to the pesticide threat, and indeed to the nature conservation movement, could shift dramatically following some incident on land or sea, or the publication of a best-seller. There were occasions when the ever-changing background to the pesticide story came to dominate the story. This book will reflect the complex interplay of cause and effect.

2

ROADSIDE SPRAYS

As its earliest Annual Reports explained, it was the task of the Nature Conservancy to seek out three types of information, namely: the biological status of individual sites, the current and future status of different plants and animals, and the effects of land use and management on wildlife. The Annual Report of 1952 cited two examples of research on changing land use: the use of chemical sprays on roadside verges and the impact of military training exercises on sand dune systems.

The large-scale use of chemicals in the environment had no precedent. There was hardly any mention of pesticides and pollution, and their effects on wildlife, in the reports and recommendations leading up to the creation of the Nature Conservancy in 1949. The reference made in the report of the Wild Life Conservation Special Committee of 1947 was prophetic but extremely brief. Noting how neither the farmer nor the conservationist welcomed the presence of pests on their land, the Committee, made up largely of biologists and ecologists, commented:

we recognize perhaps more fully the other dangers that follow upon ill-informed or indiscriminate destruction (for instance, that brought about by the improper use of insecticides) of species whose interrelations with others have not yet been ascertained.

With the exception of recently-introduced pests, such as the Colorado beetle, 'a preoccupation with the destruction of single species without due consideration of the place they occupy in their communities' might result in even more damage being inflicted on farming and other economic interests. In its report to the Minister of Town and Country Planning, the Committee recommended a more scientific approach to pest control, based on the most intimate ecological knowledge available (Wild Life Conservation Special Committee 1947).

It was only gradually that the enormity of the impact of pesticides and pollution came to be recognized. Despite premonitions, the regulation of pesticides was hardly treated as an urgent issue in the 1950s, when their significance as a form of agricultural management was becoming increasingly apparent. When the House of Commons' Select Committee on Estimates scrutinized the Conservancy in 1958, attention was focused on studies of soils and their improvement, water conservation, the effects of burning and grazing vegetation, the regeneration of woodlands, planting against erosion and flooding, and an improved understanding of animal and insect populations. Pesticides and pollution did not rank high among the priorities for research (Select Committee 1958).

For the Conservancy, the realization that a chemical revolution of major proportions was taking place began in a piecemeal manner in May 1950, when the first complaints were received of damage caused by chemical sprays to the flora of roadside verges. In 1943, British scientists had recognized the potential value of 2,4-D (dichlorophenoxyacetic acid) and MCPA (2 methyl-4 chlorophenoxyacetic acid) in selective weed control. Faced with the growing cost and scarcity of labour, the Gloucestershire County Council had been the first highway authority to make large scale use of these compounds for controlling growth on roadside verges.

The Director-General of the Nature Conservancy, Captain Cyril Diver, expressed his misgivings over this new form of verge management in a letter to the Ministry of Agriculture in July 1950. He warned of how the 'simplification of the roadside verge' could have harmful effects on both nature conservation and agriculture. The removal of plant species would deprive the fauna dependent upon them for food and shelter. The disruption caused to 'the balance of nature' might have important repercussions for the control of pests and diseases in nearby farm crops. The chemical sprays might also drift onto the crops and extend the area of damage. The Ministry's initial response was simply to argue that nothing could be done which might discourage the highway authorities from dealing with noxious weeds. Diver retorted that these species rarely grew on the verges in large numbers.[1]

The Conservancy met with greater success in its discussions with the Agriculture Research Council (ARC). For about a year, there had been increasing concern over the effects of insecticides on 'the balance of nature'. The National Institute of Agricultural Botany and the Agricultural Improvement Council had voiced their misgivings over the repercussions of hormone weed killers. On the initiative of the ARC, reference was made to these compounds at the annual conference of the Advisory Entomologists to the National Agricultural Advisory Service. The conference concluded that ecological research was required, particularly into the effects of herbicides used on roadside verges, river banks, and airfields.

The ARC's Standing Committee on Research decided to approach Captain Diver, who had also attended the conference. Although agriculturalists were vitally concerned about the effects of using herbicides, the Standing Committee felt that any research should be conducted by the Nature Conservancy. This point was emphasized by Professor J. W. Munro of Imperial College. As a follow-up to the conference, he wrote a paper outlining the benefits of a joint ARC–Nature Conservancy study.

Not only would it extend knowledge of the fauna of the hedgerow and field margin—two seriously understudied habitats—but it would provide a clearer picture of the relationship between their fauna and that of neighbouring farm crops.

Captain Diver agreed that the study of hedgerows was 'one of the most urgent projects facing the Conservancy', and appropriate for collaborative research. During his own studies of the distinctive hedgerow and wasteland species of snails, *Cepaea hortensis* and *C. nemoralis,* in the 1920s, he had been impressed by the variability and geographical range of hedgerow types, and by the way in which they represented a last stand for many plants and animals in 'our much used country'. In taking this view, Diver was supported by the Chairman of the Conservancy, Professor A. G. Tansley, and by C. S. Elton, both of whom had a considerable personal research-interest in verges and hedgerows. Elton warned of the outcry that was bound to follow, once the public discovered the threat to the roadside verges. In the year of the Festival of Britain, it was incredible that a highway engineer could change the face of England without anyone having a chance to comment.[2]

A further stimulus to concern over roadside verges came from the ARC's Unit of Experimental Agronomy, following a visit to Gloucestershire. The County Council had been so impressed by the results of spraying 4000 miles of verge in 1949–50 that a demonstration was being organised for representatives of 22 other countries. Whilst the immediate effects of 2, 4-D on open roads, where the grass verges were infested with susceptible weeds, seemed promising, the Unit of Experimental Agronomy was 'extremely uncertain' about the longer term effects, particularly where the verges were bordered by hedges. In a report to the ARC, the Unit warned that 'there is a real danger that the short-term benefits will lead to a country-wide use of the methods before the long-term effects are adequately assessed'. At the time, there was no attempt to observe changes on the treated verges in any detail, and little assurance that sufficient care was being taken in the use of the weed killers.

Right from the start, Captain Diver had recognized the need to approach the Ministry of Transport but he wanted, first of all, 'to get a united front on the scientific side'. That point was reached in April 1951, when the ARC convened a meeting of interested parties. From the chair, the Secretary of the ARC spoke of his concern that herbicides might cause cumulative injury to certain woody plants in the hedgerows and encourage weeds to become more resistant to chemical use. Such spraying programmes would make the verges harder, not easier, to manage. At the meeting, Dr A. M. Massee of the East Malling Research Station drew attention to the role of hedges as refuge for beneficial insects and Dr B. Campbell of the British Trust for Ornithology spoke of the impact of hedgerow destruction on birdlife.[3]

The key figure at the meeting was the representative of the Ministry of Agriculture, a Deputy Secretary. Because the biological evidence was so circumstantial, he doubted whether the Ministry of Transport would feel justified in discouraging spraying. He suggested that every effort should be made 'to obtain more concrete observations'. Captain Diver retorted that it was because a quick scientific answer could not be expected that some kind of administrative action should be taken in the interim period. On the understanding that the ARC and Conservancy would initiate a series of trials and surveys, the Deputy Secretary agreed to convey the misgivings of the meeting to the Ministry of Transport. In a letter, he asked the Ministry to issue a circular, recommending that the highway authorities should limit spraying for two or three years so that further investigations could be carried out.

In the words of Captain Diver, 'it now rests with us on the scientific side to provide the administrators with all the facts'. Because of the absence of any experienced entomologists on the staff (except for Diver himself), the Conservancy was particularly keen to collaborate with other bodies: a meeting of interested parties in June 1951 led to the appointment of a small team under Professor Munro. According to Diver, the main object of the short term inquiry would be 'to get the case for spraying down to realities and shorn of its advertising frills'. If spraying was the most economical form of management, it had to be given due consideration but should not be 'bolstered up by spurious arguments'. A Conservancy Research Scholarship and a special grant from the ARC made it possible to appoint two entomologists to the team. A botanist who had just joined the Conservancy, Miss Olive Balme, was seconded to the team in August 1951. The Botanical Society of the British Isles (BSBI) agreed to help find the most suitable sites for study in Gloucestershire.[4]

In corresponding with the Gloucestershire County Council, Diver emphasized that the Conservancy was concerned with the 'practical aspect, as much as the preservation of interesting and attractive flora and fauna'. The County Council pointed out that the spraying programme was only experimental and effected only 10 per cent of the roads in the country. The hormone substances were not at all poisonous to man or animal life at the levels used.

In view of the moves being made for a period of trials and observations, the Ministry of Transport agreed to issue a statement. Whilst welcoming anything to reduce the cost of verge maintenance, the Ministry indicated that it could not recommend the use of sprays 'if there were any risk of upsetting the biological balance'. In a letter to the County Councils Association, the County Councils were urged to contact the Nature Conservancy before spraying and to confine their activities for the next year or two to limited experiments. By then, 'it should be possible to judge whether the dangers that are foreseen can be reasonably ignored or prevented'.[5]

Trials and observations

Up to this time, all the available expertize and experience in roadside spraying had resided with the company that had a virtual monopoly over the supply and application of sprays. Professor Munro argued that the time had come to appraise these methods independently and, in his various reports, he called for closer collaboration with the Road Research Laboratory. Following a meeting of interested parties in January 1953, the Ministry of Transport agreed to provide funds for the Laboratory to investigate the economic aspects of spraying. The Gloucestershire County Council and its contractors agreed to cooperate in the investigations.

During 1953, a collaborative study was mounted on 14 sites, with a total length of 40 miles, in Gloucestershire. In answer to a Parliamentary Question in July 1954, the Minister of Transport drew attention to the restraint being shown by the highway authorities in their use of chemicals and to the research being carried out in collaboration with the County Councils. In an interim report of April 1955, the Road Research Laboratory indicated that a single spraying once a year in spring might be sufficient to keep the vegetation low without eliminating it. This could lead to a 55 per cent reduction in man-power requirements.[6]

The grants awarded by the ARC and Conservancy ended in 1954. In a final report, Professor Munro described how the population of 'visiting insects' reflected the use of neighbouring land, rather than the condition of the roadside verge. Thus, the hover fly was most common on those verges adjacent to cattle-grazed pastures. The insect fauna of Umbelliferae occurred widely and was not dependent on the verges. In presenting these and other results, Munro stressed the severe limitations of a short term study. The insect fauna of the roadsides was part and parcel of the fauna of the countryside and it was therefore both unreal and unscientific to regard the verge as a unique, isolated habitat. He also complained of the difficulties of getting insects identified, of discovering their habits and life histories, and of understanding host–parasite relationships.[7]

The incomplete nature of the scientific evidence and the loose way in which the various parties had collaborated made it especially difficult to arrive at a consensus in interpreting the results of the three-year study. Professor Munro concluded that, although spraying would have a marked effect on the insects living on the verges, there was little or no evidence that the populations as a whole would be affected to any appreciable extent. The Conservancy believed such a judgement was premature and certainly should not be taken as an excuse for the resumption of unrestricted spraying. Most counties had interpreted the Ministry's letter to the County Councils Association as a recommendation to stop all spraying.

The Conservancy was reluctant to concede any relaxation, particularly in view of the lack of any well-defined objectives for spraying. The use of sprays against noxious weeds was based on false assumptions and their deployment for tidying up margins and improving visability around blind corners could easily become an excuse for destroying the whole habitat. Charles Elton described the verges as 'one of our most important nature reserves'. Apart from bird protection, the Conservancy should make the protection of the roadside verges the subject of its 'first determined campaign'.[8]

In its discussions with the Ministry of Transport, the Conservancy was able to draw on the results of trials conducted by Miss Olive Balme, the botanist in Munro's team. There was a strong possibility that the wildlife interest of the verges would be eliminated and that the highway authorities would fail to reduce the long term expense of managing the verges.[9] As Miss Balme explained, in papers given to the British Weed Control Conferences of 1954 and 1956, a single application in early spring might be sufficient to eliminate such species as meadow cranesbill (*Geranium pratense*), creeping buttercup (*Ranunculus repens*), dandelion (*Taraxacum officinale*), and woundwort (*Stachys sylvatica*), but there was evidence that the effects on the most conspicuous weeds would be only temporary (Table 2.1). Some of these, killed by spraying, were likely to recolonize locally-disturbed areas so that lasting control could only be achieved through repetitive spraying. In the particular case of biennials, such as cow parsley (*Anthriscus sylvestris*) and hogweed (*Heracleum sphondylium*), the flowering shoots might be killed, but spraying usually failed to reach the first-year vegetative growth, or the underground root-offsets, many of which were already separated from the parent plants. Thus growth in the following year would be unimpaired and the population could maintain itself from year to year, despite spraying (Balme 1954, 1956).

Circular 718

It was against this background of trials and observations that the Conservancy and the Ministry of Transport reached an agreement in 1955 on how spraying might be regulated in the future. The highway authorities would be told that the Conservancy did not object to the use of selective sprays of the phenoxyacetic-acid type if four conditions were met. First spraying should only take place on trunk and class A roads, and on dang-

Table 2.1 The susceptibility of species of spraying treatments (after Balme 1954)

Species showing significant reductions, with controls	Species of doubtful susceptibility (5–20%)	Species showing no differential counts between treatments and controls
Fragaria vesca	*Angelica sylvestris*	*Anthriscus sylvestris*
Geranium pratense	*Achillea millefolium*	*Arum maculatum*
Galium mollugo	*Arctium lappa*	*Crataegus monogyna*
Glechoma hederacea	*Agrimonia eupatoria*	*Galium aparine*
Hedera helix	*Cornus sanguinea*	*Geum urbanum*
Lathyrus pratensis	*Cirsium arvense*	*Geranium robertianum*
Pastinaca sativa	*Galium cruciata*	*Heracleum sphondylium*
Plantago lanceolata	*G. verum*	*Lamium album*
Potentilla reptans	*Rumex crispus*	*Lapsana communis*
Prunus spinosa	*Viola* spp.	*Mercurialis perennis*
Ranunculus repens		*Ranunculus ficaria*
Stachys sylvatica		*Rubus fruticosus*
Taraxacum officinale		*Scilla non-scripta*
Urtica dioica		*Veronica chamaedrys*
Vicia sepium		

erous corners on class B roads, at the earliest possible stage of growth. Secondly, only a width of 10 feet from the roadside edge should be sprayed, and every care was to be taken to avoid drift or harm to adjacent hedgerows, crops, and gardens. Thirdly, the highway authorities should not spray sections known to have interesting species or communities present. Fourthly, spraying was still in an experimental phase and further changes could be made in the light of experience. The substance of the agreement was set out in a circular which the Ministry of Transport issued to the highway authorities as Circular 718 in August 1955.[10]

It was one thing to reach an agreement in London and to publish a circular; it was another to ensure that the agreement was implemented at the local level. By the autumn of 1957, the Conservancy had received many complaints of indiscriminate spraying from the BSBI and from other naturalists' bodies, particularly from Staffordshire where the county surveyor argued that he no longer had the labour available to manage the verges in the traditional way. Following representations from the Conservancy, the County Council proposed a compromise which the Conservancy rejected on the grounds that it still ran counter to the national agreement.[11]

The Ministry of Transport was quick to point out that the circular was not an instruction. Not only could the highway authorities ignore the Minstry's advice but they had complete jurisdiction over class C and unclassified roads. The Conservancy admitted that this was the 'legal' position, but the Ministry had nevertheless entered into 'a gentleman's agreement'. If the Ministry had known the agreement could not be made effective, why had it negotiated it in the first place? The Conservancy had made major concessions; Circular 718 embodied the minimum requirements for wildlife. Far from continuing to restrain critics, the Conservancy would join the voluntary bodies in attacking the way in which the highway authorities were flouting a scientifically based policy.[12]

After further discussions with the Conservancy, the Ministry wrote to the Staffordshire County Council, drawing its attention to the representations of the Conservancy. There was no doubt that 'powerful national organisations' were concerned over the effects of spraying on the amenity of the countryside and that the losers in any conflict 'would undoubtedly be the highway authorities'. Having considered the various representations, the County Council decided in July 1958 to continue its spraying programme, excluding from it 211 miles of road that had been the subject of special representations.

Both the Ministry and Conservancy were anxious to resolve the conflict. The right tactics had to be adopted. If there were any suggestion of Whitehall attempting to override local independence, it was likely that the local people would win the day. At the Ministry's instigation, a meeting took place between the Clerk of the County Council and senior staff of the Conservancy. The Director-General stressed that the Conservancy was anxious to avoid 'dog-fights' with individual authorities. Some members of the Conservancy, including the Chairman, were also members of highway authorities; they appreciated the peculiar difficulties faced by Staffordshire, but could not overlook the fact that what happened there was a test case for the effectiveness of Circular 718.

In March 1959, the County Council announced that spraying would continue because of the exceptionally heavy road-works programme and the severe competition for labour from industry. Spraying would, however, be confined to autumn, when the flowers had died, the crops were harvested, and the birds were no longer nesting. Any stretches of exceptional beauty or scientific interest would be excluded. The Conservancy would be told when and where spraying was taking place. These concessions were still regarded as inadequate, but the Conservancy welcomed the softening of attitude and had no practical alternative but to accede. The task of organizing naturalists to observe the effects of spraying would provide the Conservancy with valuable experience of surveillance work, carried out by voluntary bodies.[13]

Staffordshire was not the only county to arouse concern. Even where highway authorites complied with the Circular, instances of indiscriminate spraying still occurred. During a conference sponsored by the Society for the Promotion of Nature Reserves in May 1960, delegates from the various conservation and natural history bodies were taken to see the roadside verges of Tetford Hill in Lindsey. In a letter to *The Times*, the Secretary of the BSBI, J. E. Lousley, described how they were astounded to find all the plants blackened and distorted by recent spraying. Not only was the verge a Site of Special Scientific Interest but the Lincoln-shire Naturalists' Trust had stressed its importance to the County Council on numerous occasions.

No roadside verge was safe from spraying. The Lindsey County Coun-cil had an exceptional record for its general interest in nature conservation and it immediately apologized for the incident. Its opera-tors had acted without instructions, there had been a change of personnel, and details of the natural history interest had been misfiled. The Council once again informed its operators of the contents of Circu-lar 718 and it offered to erect markers to identify those verges of high biological interest. Whatever precautions were taken, there was no guar-antee that such incidents would not occur again, whether in Lindsey or elsewhere.[14]

Instead of relying on chance discoveries and unsolicited reports, the time had come to appraise the effectiveness of Circular 718. All the Con-servancy's regional offices were asked to report on the position in their respective areas—the outcome was a considerable shock for the Conser-vancy. The only substantiated cases of abuse were reported from Staffordshire, Lindsey, and a class B road in Monmouthshire. Either the regional offices were out of touch or many reports of indiscriminate spraying had been exaggerated. Either way, the Conservancy was clearly not in a position to demand action from the Ministry. All further com-plaints would have to be confirmed by the regional staff so that a true picture of the exact position could be obtained. At a liaison meeting with the voluntary bodies, the Conservancy warned of how the case against indiscriminate spraying would be undermined in the eyes of the Ministry if any complaints proved groundless.[15]

A comprehensive 'policing' of the use of herbicides was impossible. The divisional road engineers and their staff lacked any biological train-ing and the Conservancy's regional staff were thin on the ground and hard-pressed with other duties. In an attempt to improve understanding, the Conservancy convened a meeting on the technical and scientific aspects of roadside management in November 1963. The principal prop-osal to emerge was a suggestion from the County Surveyor for Gloucestershire that spraying should be permitted on *all* roads, up to a width of 8 feet from the roadside. He argued that the classification of roads was irrelevant to nature conservation and that the overall effect would be to create a transition from short to long vegetation on all verges. The Conservancy agreed to the proposal in principle, subject to three important constraints. Where possible, the verges should be cut, rather than sprayed; verges of high biological importance should remain untreated; and not every verge in a district should be cut or sprayed at the same time.[16]

The Ministry of Transport believed the county surveyors would be more sympathetic if the scientific role of the verges was explained more fully to them. The Conservancy took up the suggestion, and a paper. 'Roadside verges—their significance for biology and conservation', was drafted by Dr Norman Moore. It explained the increasing value of the verges as a reservoir for wildlife, and how this focus of scientific interest might be harmed by spraying. The Ministry objected that a paper of this type would cut absolutely no ice with the surveyors. To succeed, it had to demonstrate the economic advantages of conserving the wildlife of the verges or identify the kind of tangible national disaster that would arise in the absence of such a policy. The Conservancy suspected that the Ministry was baulking at the prospect of being the first to appeal to the aesthetic sensibilities of the surveyors.[17]

Throughout its discussions with the Ministry and the voluntary bodies, the Conservancy was aware that its recommendations were based on very little 'hard' evidence. The Annual Report of 1955 expressed the hope of extending the studies initiated by Miss Balme, but this did not become a real possibility until the 1960s. By that time, rotary and flail mowers were beginning to displace chemical forms of control. Frequent cutting by these methods was capable of producing an equally dull, uniform grass-cover. Experimental plots were laid out in 1965 in order to compare the effects of applications of 2,4-D with mechanical treatments.[18] Whereas the trials of a decade earlier had been an *ad hoc* response to an emerging problem in land management, these new studies were a logical extension of the Conservancy's far-reaching studies of the effects of pesticides and pollution on the natural environment.

Many years later, Nicholson (1970) described the discussions and studies leading up to Circular 718 in 1955 as being of considerable importance, in so far as they 'served notice on commercially interested groups, and on public administrative and technical officers susceptible to their propaganda, that the application of these dangerous substances would in future be carefully watched and strictly supervised in the overall public interest'. The introduction of the Circular and the practical difficulies of enforcing its recommendations had a significant, albeit indirect, bearing on the conduct of the agricultural pesticides debate. The Conservancy entered that debate conscious of some of the administrative and scientific problems that were likely to arise.

3

THE CHEMICAL REVOLUTION

There have always been instances of localized pollution. Mesopotamian mythology recounts how earthquakes in the Zagros mountains released crude oil and burning natural gas into the environment (Kinnier-Wilson 1979). Many examples of pollution arising from human activities can also be cited. It is the scale, rather than the incidence, of pollution that is so striking today. Many living organisms are being affected significantly for the first time in their evolutionary history (Moore 1970*a*). The extensive effects of this change in the scale of pollution are graphically illustrated by the story of pests and pesticide control—a story in which the Nature Conservancy has played a conspicuous part.

For many generations, farmers sought the perfect pesticide which would kill pests but which would leave crops and livestock unharmed. In an essay on weeds in agriculture (Holdich 1825) it was claimed that common salt, dropped on the crown of a weed, would destroy the plant but leave the grass uninjured. Most pesticides proved to be ineffective, not sufficiently selective, or too expensive in relation to returns from the crop. The utility of the insecticide, pyrethrum, was severely limited by the cost and difficulty of growing and harvesting the plant from which it was extracted, *Pyrethrum cinerariaefolium*. Supplies were always expensive and often inadequate.

Before the last war, the only pesticides in common use were either derived from plants (such as pyrethrum, nicotine, and derris), or were simple inorganic chemicals (such as arsenate of lead). Mercury or mercurial fungicides were first used in Britain in 1929, and over a million acres of land were treated each year throughout the 1930s. The only important synthetic organic chemical, dinitro-ortho-cresol (DNOC), was introduced as a weed killer in 1932. During the war, the synthetic plant-growth regulators, MPCA and 2,4-D, were discovered independently in Britain (see page 7) and America. A new era in chemical control had begun. Absorbed by the leaves the herbicides passed to other parts of the sprayed plants. Because of differences between plants in regard to the retention, penetration, trans-location, and cell-level toxicity of the compounds, it became possible to kill weeds chemically without damaging the crops (Fryer 1964).

Meanwhile, the organochlorine (OC) insecticides, DDT and gamma-BHC, had been introduced, (later known as HCH). In Germany, organophosphorous (OP) compounds were developed as potential nerve gases. Known as organic compounds, with a molecule based on a

16

skeleton of carbon atoms, the former compounds contained chlorine atoms, while the latter had a phosphorous atom in the centre of the molecule. For many, DDT (dichloro-diphenyl-trichloro-ethane) had all the virtues of an outstanding insecticide because of its toxicity to a wide range of insect pests, low mammalian toxicity under normal circumstances, and unprecedented persistence.

For many years, chemists had sought a synthetic contact insecticide, capable of being made in the laboratory and manufactured on a large scale in chemical works, which would rival such vegetable insecticides as pyrethrum in effectiveness and surpass them in cheapness and availability. DDT had first been synthesized in 1874, but no one had found a practical use for it until 1940, when a research group in the Swiss firm, J. R. Geigy AG, discovered how effective the compound was in controlling potato beetles and clothes moths. Samples were forwarded to American and British representatives of the company in 1942. In a paper of 1945, the Director of the ARC Unit of Insect Physiology in Britain, Dr V. B. Wigglesworth, recalled how he and his colleagues had been extremely sceptical of the claims made for the new insecticide. The patents taken out by the company 'looked too good to be true'. In the event, the trials carried out in America and Britain not only substantiated the claims but led to further uses for DDT being found (Wigglesworth 1945). By the middle of 1944, DDT was standard issue for checking the spread of typhus, malaria, and other killer diseases among Allied Forces and civilian populations in every theatre of war. By the end of the war, so much DDT and equipment for applying the insecticide had been manufactured that some could be spared for veterinary purposes and for general use in farming and forestry (Dunlap 1981).

DDT entered an animal's body either through the external body surface or, if present in its food, through the gut wall. Either way, the pesticide was carried round the body by the blood or circulatory system. Each molecule of DDT then took one of four possible pathways (Fig. 3.1). Like other pollutants, the compound could kill an organism if sufficient molecules, avoiding metabolism, storage, and excretion, reached the site of action, to set in motion a chain of events that would disrupt those physiological activities essential to life. Many years were to elapse before the significance of each of the pathways within, and between, organisms became clear, both in terms of the target species and the wider environment (Moriarty 1975b).

In response to demands for greater food production, farmers demanded more and better pesticides. It was estimated in the United States that a tenth of the annual harvest was destroyed by insects. Farmers had heard of the way in which DDT had killed lice, flies, fleas, and mosquitoes, thereby saving the lives of millions of military and civilian

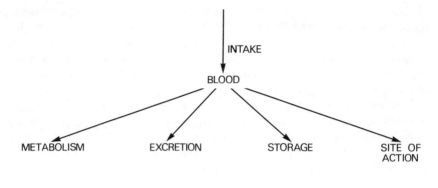

FOUR POSSIBLE PATHWAYS FOR
EACH MOLECULE OF DDT

Fig. 3.1 Four possible pathways for each molecule of DDT in an organism.

personnel during the war, and they demanded the application of these great strides in organic chemistry to the needs of agriculture. They were not disappointed. A multi-million dollar industry grew up almost overnight; the United States soon led the world in both production and consumption of pesticides.

The 'discovery' of DDT had provided a tremendous stimulus to the search for the universal insecticide, the compound that would kill every insect with which it came in contact, without harming the plants on which it was spread or causing injury to human beings eating those plants or coming in contact with the compound during its manufacture or application. The trials carried out during the war suggested that DDT, used with adequate care, posed no threat to human beings or to domestic and farm animals. Over the previous half century, a great deal of experience had been gained in the marketing and regulation of pesticides in the United States. There was already an infrastructure in existence, ready to develop and promote DDT and the new pesticides (Dunlap 1981).

Albeit on a much smaller scale, the increased use of pesticides in Britain was equally dramatic. The scientific and commerical origins of the industry have been highlighted by a history of one of the pioneering companies, Pest Control Ltd. Right from the start, in 1939, the company's founders, Sir Guy Marshall and Dr Walter Ripper, envisaged:

a worldwide 'entomological industry', with research capable of solving any pest control problem, chemical plant producing the necessary chemical sprays, engineers making the required special spraying machines, and a contract service poised to roll into the fields.

Although raw materials were in short supply, the war stimulated food production greatly. The Ministry of Agriculture and local advisers urged farmers to use pesticides on an increasing scale. The rise in farm profits and the concurrent decline in labour supply provided both the scope and need for contract spraying. By 1944, the company's 'Hints for higher harvests' offered 20 contract services (Anon. 1976).

By the mid-1960s, about 200 pesticides were in common use. Because 90 per cent were synthetic organic compounds rather than naural products, they could be produced in unprecedented quantities (Woodford 1964). Herbicides were particularly welcome at a time when labour costs had made hand hoeing so expensive, and when the farmer wanted to concentrate more and more on cereal production. It was a time when housewives were also demanding better-quality produce. Apples had to be free from blemishes, and potatoes free from holes. Although the relationship could not be measured precisely, there was little doubt that pesticides made a major contribution to the post-war increase in efficiency and productivity in farming. Table 3.1 correlates the rise in yields of various crops, as indicated by the Annual Returns to the Ministry of Agriculture, with the estimated increase in the costs and benefits of insecticide treatments (Moore 1971). It was estimated that chemicals saved over £100 million annually in reduced weed competition and pest depradation (MacGregor 1964), and Strickland (1966) believed insecticides alone save farmers an estimated £25 millions a year in reduced damage.

Table 3.1 Increase in yields and the value of insecticides, 1945–66

Crop	Yield per acre (England and Wales) in hundredweights		Annual value (£ million) at average 1959–64 prices of crops believed saved by insecticides in England and Wales	Annual approximate cost (£ million) of insecticide treatments believed applied in England and Wales
	1945	1966		
Wheat	19.0	30.5 ⎫		
Barley	19.0	28.2 ⎬ 6.71		0.99
Oats	17.6	27.6 ⎭		
Sugar beet	188.0	294.0	3.37	0.26
Potatoes	144.0	202.0	2.56	0.37

The Zuckerman Working Parties

Formidable obstacles were laid in the path of those calling for a more moderate use of pesticides. The dangers arising from DDT and other organic compounds seemed very remote and slight in relation to the obvious superiority of the new pesticides over lead arsenate. For the pesticide and farming industries, it was merely a case of including the new generation of pesticides within the existing framework of regulations which had evolved over the previous 50 years, based largely on unformulated and almost unconscious assumptions. It was some time before the users of the new pesticides realized that these regulations would no longer suffice and that the new substances were much more powerful than even they had realized, or been prepared to accept (Dunlap 1981).

It is not difficult to find early premonitions of the damage pesticides might inflict on wildlife. In his paper, published in *Atlantic Monthly* in 1945, Wigglesworth outlined not only the merits of DDT but also the manner in which it might upset 'the balance of nature'. Published in the magazine's series, 'A scientist looks at tomorrow', the paper warned that DDT might act 'like a blunderbuss discharging shot in a manner so haphazard that friend and foe alike are killed'. The broadcasting of sprays from the air was likely to kill 'vast numbers of insects of all sorts' and, if carried out on an extensive scale, there would be little chance of recolonization. The experience of the previous 30 years had indicated that insecticides could be a two-edged sword, upsetting the natural balance between insect enemies and friends. Moreover, because in time species often developed a resistance to the compounds, more and more pesticide might have to be applied to achieve the same results.

No one denied the need for insecticides but, as Wigglesworth (1945) emphasized, it was important to apply them at such a time and in such a manner as to cause the greatest damage to pests and the least injury to their natural enemies. In the long run, it might be better to use an insecticide that killed 50 per cent of the pest insects and none of its predators or parasites, rather than another which would kill 95 per cent of the pest and all its natural enemies. In the search for a more selective type of pesticide, it was essential to discover more about 'the interaction of pests with their physical environment and with the other insects around them'. In Wigglesworth's words, 'we need to know far more about their ecology—that is, about their natural history studied scientifically'. Wigglesworth's paper was followed by others, particularly in the United States. The *Audubon Magazine* published 30 articles and editorials on the subject of DDT over the period 1945 to 1962.

The first discussion of the problem at an international level took place at the International Technical Conference on the Protection of Nature in August 1949. In a review of the effects of insecticides on the balance of nature, five types of side-effect were identified, namely:

1. Insecticides might directly kill non-target species.

2. They might endanger insectivores feeding on insects affected by pesticides.
3. The large-scale destruction of insects and their larvae might cause insect-eating species to starve or emigrate.
4. Pollination might consequently be jeopardized.
5. Insecticides might disrupt normal predator–prey relationships.

Delegates at the international meeting recommended that such bodies as the Food and Agriculture Organization, World Health Organization, and UNESCO, should set up a Joint Permanent Commission on Pesticides. They warned that the side-effects of pesticides might become so significant as to counterbalance the benefits of spraying. Pest damage might even become worse. Pesticides might kill not only the target species but also predatory species which had a lower fecundity and, therefore, a much lower potential for recovery from the effects of spraying. Delegates set out ways in which the ecologist could assist in reducing or minimizing the hazardous and unintended effects of modern chemicals (Anon. 1950).

Despite these admirable sentiments, some delegates appeared to be extremely complacent in their attitude towards pesticide use (Nicholson 1970). A representative of the Department of Agriculture in Washington assured the Conference that 'there were enough experts at the disposal of the US Government to fulfil the task of the controlled use of pesticides'. Five years later, a representative of the Department of Agriculture was equally reassuring in his address to the Fifth Technical Meeting of the International Union for the Protection of Nature. He said that all new insecticides were thoroughly tested by entomologists, soil and plant scientists, and toxicologists employed by the manufacturers, and before being registered with the United States Department of Agriculture.

Not everyone was persuaded by such assurances. A spokesman for the US Fish and Wildlife Service described how the rapid appearance of new formulations of pesticide outstripped every attempt to discover the compounds and methods of application which would cause least harm to wildlife. The Wildlife Service could afford to employ only one full-time scientist on the problem. The Meeting of 1954 contributed further resolutions to those of 1949. In order to avoid damage to beneficial and harmless forms of insect, pesticides should be made as specific as possible, applied in a way which would cause least danger to other wildlife, and always be used in accordance with any recommendations made by official scientific bodies (Anon. 1956).

It would be very misleading to suggest that these recommendations aroused much interest or sympathy. Much more concern was shown in Britain about the effects of pesticides on human health. For many years,

DNOC had been applied to fruit trees in early spring; trouble began when the compound was used in much higher concentrations as a herbicide on cereal crops. Seven workers died of DNOC poisoning between 1946 and 1950. The highly toxic organophosphorous insecticides were also being used on an increasing scale. Over 200 workers were reported to have died of their side effects in the United States. The dangers were likely to increase in Britain as pest control came to be carried out on a large scale by contractors, whose operators would be exposed to risks over much longer periods of time.

Whilst experts from various countries might exchange experiences, and sound warnings at international gatherings, any response to the perceived threats to man and other organisms had to take into account of the history and philosophy of individual countries in respect to legislative and other methods of regulation. In Britain, the earliest initiatives arose from a concurrent concern about safety in the workplace and the incidence of food adulteration (Martin 1963).

In 1949, an official committee on Health, Welfare, and Safety in Non-industrial Employment (the Gowers Committee) recommended that employers should provide protective clothing whenever poisonous sprays were being used (Secretary of State 1949). This was endorsed by a Working Party of leading scientists, set up by the Ministry of Agriculture to investigate the problem further (Working Party 1951). Under the chairmanship of Professor Solly Zuckerman, the Working Party recommended statutory backing for those safety measures which many employers were already implementing on a voluntary basis. The Government agreed, and an Agriculture (Poisonous Substances) Bill was passed without opposition in 1952.[1]

By the time the report of the Working Party was published in 1951, there was mounting concern on two other counts, namely the possibility of people being harmed by pesticide residues in their food and the effects of pesticides on wildlife. Both were of major significance. Professor Zuckerman decided that 'the consumer aspect was the more urgent and should be dealt with first', and his Working Party was reconstituted for that purpose.[2] It was a decision of considerable importance for nature conservation. When the time came to consider the wildlife aspects, account had to be taken of a further set of pesticide regulations. Whilst consumer safeguards might benefit wildlife indirectly, they might also deter the Government from contemplating a further set of controls imposed explicitly on behalf of wildlife.

The reconstituted Working Party issued its report in 1953. Whereas the study of hazards to operatives had taken only six months, the study of treated foodstuffs took two years. The issues were much less clear-cut. Little was known about the extent of pesticide use on specific food crops or about the actual levels of residue in different foodstuffs. The toxico-logical significance of any residue found was often unknown, and methods of analysis for detecting them were either lacking or could not be adapted for enforcement purposes (Working Party 1953; Moore, W. C. 1964).

The lack of scientific data and the complexity of the issue may have been the most important reasons for the Working Party not recommend-ing any further legislation. It suggested that a voluntary system of control should be adopted, whereby all interested parties, including the chemical manufacturers, would cooperate in minimizing any harmful side-effects. They would do so by conforming with a notification procedure adminis-tered by the Ministry of Agriculture, Fisheries and Food, on behalf of all interested Government departments. All manufacturers would be expected to observe the conditions laid down by the scheme, participa-tion in which would be free of charge. In the event of the scheme not being effective, the report of the Working Party made it clear that legisla-tion would have to follow (Miller 1965). As recommended, an Inter-departmental Advisory Committee on Poisonous Substances used in Agriculture and Food Storage was set up in 1954 in order to keep the incidence of residues in human foodstuffs under close surveillance. One of its first acts was to appoint an advisory Scientific sub-committee.

Meanwhile, there was correspondence in *The Times* about the effects of pesticides on game and wildlife. The British Field Sports Society con-vened a meeting in October 1952 at which resolutions were passed expressing alarm over the harmful effects on animal and plant life, and calling for immediate curbs on the use of the more harmful sprays. Dur-ing the autumn of 1952, there had been an unprecedented number of reports of birds and mammals being found poisoned. There had been an unusually late and heavy attack of cabbage aphid on Brussels sprouts, and growers had responded by spraying their crops with schradan, a sys-temic organophosphorous compound which had not previously been used on so extensive a scale. Because of the lateness of the season, the sprouts were the only tall-standing crop available as covert. This combi-nation of factors led to many reports of birds and other animals being found dead in, or near, the treated crops. One such incident, recorded by the ICI Game Research Station, affected 46 acres of Brussel sprouts at Charlton Abbots in Gloucestershire (Table 3.2). Because a complete search of the area was impossible, the figures in the table underestimated the numbers killed (Middleton 1956).

The attitude of the agricultural bodies to such incidents was summed up by a representative of the Agricultural Research Council, who wrote that 'everyone is, I'm sure, anxious to see safer materials employed, but until they are forthcoming there is little that can be done'. He explained

Table 3.2 Birds and mammals found dead after spraying with schradan

Animal	Days 1st	2nd	3rd	4th	5th	6th	Total
Partridge	—	10	4	2	3	1	20
Pheasant	—	2	4	3	—	1	10
Blackbird	9	13	8	2	—	—	32
Thrush	4	7	2	1	—	—	14
Chaffinch	6	16	7	1	—	—	30
Bullfinch	—	2	—	—	—	—	2
Tit	3	8	4	2	—	—	17
Linnet	2	5	4	—	—	—	11
Yellowhammer	1	6	1	—	—	—	8
Dove	1	1	—	—	—	—	2
Sparrow	2	7	3	—	—	—	12
Robin	—	—	1	—	—	—	1
Rabbit	—	—	2	1	2	2	7
Hare	—	—	1	1	—	—	2
Rat	—	—	1	1	—	—	2
Mouse	—	1	2	1	—	—	4
Squirrel	—	—	—	—	1	—	1
Stoat	—	—	—	1	—	—	1
							Total 176

that the Zuckerman Working Party had rejected the withdrawal of insecticides as a means of avoiding hazards to operatives, and that 'the evidence of damage to birds, though indisputable, is not extensive and would not at the moment justify such a radical procedure as that of prohibition'. In reply to a Parliamentary Question proposing the banning of the use of DNOC without licence, the Minister of Agriculture argued that he had no power to require notification of spraying. He reiterated that the long-term solution lay in producing equally effective sprays that were not toxic to wildlife.[3]

The attitude of the Nature Conservancy was set out by the Director-General, Cyril Diver, in a letter to the Ministry in April 1951. There had been a rapid increase in the 'improper, unwise and unskilled use' of toxic sprays. Contract spraying was a profitable business, and operators needed no qualifications. The Conservancy believed that uncontrolled use of the chemicals might be as harmful in the long-term as their proper use could be beneficial. The Ministry of Agriculture was, of course, quick to point out that there was no convincing evidence to support such a contention. As Diver conceded, it was another case of the forces of destruction advancing far more rapidly than knowledge of the consequences of destruction. It took longer to collect that kind of information but, Diver warned, unless something was done quickly, irreparable damage would be done. For its part, the Conservancy hoped to initiate relevant studies on hedgerows, verges, and roughs (see p. 8), but, as a new organization, the Conservancy was handicapped by difficulties in finding staff to undertake and direct research, and by cuts in its budget. Even if adequate research programmes could be mounted, there would be no quick scientific answers.

As work on its second report neared completion in 1952, preparations were made for the Zuckerman Working Party to take up the wildlife question. Cyril Diver welcomed the further reconstitution of the Working Party, and asked that the Conservancy should be represented. In a discussion on the new terms of reference, it was suggested that the use of toxic chemicals by highway authorities should be included, and that reference might be made to hormone weed killers. This was rejected by one of the two secretaries of the Working Party on the grounds that it would 'extend the scope of the inquiry into a much more complicated field than we had intended at this stage'. In the end, it was agreed that Professor Zuckerman should be invited informally to look at the question of hormone weed killers and to make whatever reference might be necessary in the eventual report.[4]

The formal terms of reference of the Third Zuckerman Working Party were:

to investigate the possible risks to the natural flora and fauna of the countryside from the use in agriculture of toxic substances, including the possibly harmful effects for agriculture and fisheries of the destruction of wild life.

The 17 members of the Working Party each represented some aspect of agriculture or fisheries. The only exceptions were the Chairman, and the representatives of the Medical Research Council and Nature Conservancy. The latter was represented by Captain Diver, who had by then just retired as Director-General, and by Sir Norman Kinnear, a member of the Conservancy and of the Home Office Wild Birds Advisory Committee (Working Party 1955).

In its report of June 1955, the Working Party stressed that there was no such thing as 'fixed balance of nature'. There was, however, evidence that changes were taking place on a far more extensive and intensive scale than ever before, and the question had to be asked whether such innovations as chemical pesticides would have overall, a beneficial effect. Before giving them a general blessing, there had to be an assurance that pesticides would do more good than harm, both in the short and longer term. To this end, the Working Party decided to pursue its inquiry under four headings, namely:

1. Which toxic chemicals used in agriculture constituted a danger to plants and wild animals.
2. In what circumstances did the dangers arise
3. Were there any precautionary measures by which these dangers could be obviated or minimized?
4. Were additional legislative powers needed?

The amount of relevant information submitted to the Working Party was slight and inconclusive. After only two meetings, the Chairman wrote to the Minister of Agriculture that it was clear that the 'intermittent public outcry against the use of these chemicals' could only be assuaged if it could be proved that either 'the danger is not a real one or that we have already framed measures to minimize the casualties'. The only way the Working Party could obtain that proof was for it to organize field inquiries and experiments. The Minister gave his consent, and a sub-committee was set up to devise criteria for a series of trials on DNOC and schradan. The sub-committee was made up of personnel from the Conservancy and the ICI Game Service. The Conservancy agreed to meet the costs of any staff involved in the trials; the Ministry would defray all other expenses.[5]

As early as February 1953, the Conservancy had recognized the need for such studies. The regional officer for East Anglia, Dr Eric Duffey, was asked by the Conservancy's new Director-General, E. Max Nicholson, to take charge of:

a small-scale research programme directed to ascertaining exactly what happens in the field when spraying takes place and to get keepers, etc., in sample areas to pick up all the birds and mammals affected so that the damage could be correlated with the seasonal and weather conditions, the strength of the dosage used in spraying and so forth.

It proved to be a frustrating assignment. Whereas it was possible to organize trials on roadside verges through the aegis of the county councils and Ministry of Transport, every landowner and farmer had to be approached in any study of farmland.[6] Moreover Dr Duffey's plans to study the effects of organophosphorous sprays, which were used against the yellow virus disease on sugar beet, were upset when the aphid carrier failed to materialize and require spraying, and his request to see the complaints file of the major pesticide firm in the region met with a refusal to co-operate. Despite such setbacks, he was able to suggest that the threat to wildlife from pesticides seemed to be small. A laborious search over a hundred acres of crops sprayed with DNOC revealed only one dead bird, a pheasant. Direct contact with DNOC did not appear to cause much harm to wildlife, judging from the number of bright yellow hares and game birds observed! Duffey warned, however, that the problem was so complex that it ought to be more fully studied by a university team or at least by a full-time worker.[7]

The direct involvement of the Zuckerman Working Party suggested the possibility of a more ambitious census being taken of wildlife, both before and after spraying had been carried out. In the event, only two studies on the effects of DNOC were mounted in 1954; one in mid-April, at West Barsham in Norfolk, under Dr Duffey and the other in early May, at Lockinge in Berkshire, under Dr Norman Moore, the regional officer for south-west England. Although both studies provided useful experience in bird census work, no firm conclusions on the effects of spraying could be deduced from the results. In his report, Duffey attributed this to the fact that the Norfolk area was already impoverished of birdlife, and the experiment had been carried out when the sprayed crops afforded little food or cover for wildlife. The trial with schradan was no more successful. Only 37 acres were treated, and heavy rain fell soon after spraying, diluting the spray or washing it into the soil before absorption.[8]

In the absence of firm evidence from this or any other quarter, the Working Party could only make general observations and recommendations in its report. In normal circumstances, there appeared to be little chance of birds and mammals inhaling lethal doses of pesticides or absorbing sufficient amounts through their skin to cause death. Deaths occurred largely as a result of ingesting toxic chemicals on crops or within prey already contaminated. Ingestion was only likely to occur within three or four days of spraying. The most harmful sprays were likely to be:

(1) organophosphorus insecticides applied to brassicas in late summer;
(2) arsenical compounds used for potato-haulm destruction in September;
(3) dinitro weedkillers applied to corn and peas in spring and July;
(4) DDT insecticides applied to orchards, carrots, and peas.

Not only should the chemicals be handled with greater care, but steps should be taken to secure a better understanding of how individual forms of wildlife were affected by spraying. The only chemicals to be used should be those effective against pests, and which caused the least possible damage to wildlife. Greater restraints should be exercised over the use of dusts and arsenical mixtures. All chemicals should be applied in the lowest concentrations practicable, and at times when the spray would dry quickly on the plant. In order to avoid the spray drifting onto other land, booms should be covered and not used on windy days. Every effort should be made to flush game and wild birds and mammals from the areas to be treated. Care should be taken not to spray watercourses or ponds, and not to dump pesticides or used containers about the country-side.

The Working Party did not believe it was necessary or practical to enforce these precautions by laws. It recommended instead that the membership of the Inter-departmental Advisory Committee on Poison-ous Substances used in Agriculture and Food Storage should be widened to include representatives of nature conservation interests. The terms of reference of that Committee should be widened:

(1) to receive and consider evidence and reports of research on the risks to wildlife resulting from the introduction of new chemicals;
(2) to bring to the notice of appropriate bodies any need to investigate the effect of toxic chemicals;
(3) to disseminate information;
(4) to advise on the need for legislation or new regulations.

The Minister of Agriculture accepted the recommendation, and two officers of the Conservancy were added to the Committee in September 1955. They were Mr Robert E. Boote and Miss Olive Balme, who was conducting the trials on roadside verges (see p. 11). Boote had been appointed as a Principal to the Nature Conservancy in 1954 and he had soon taken charge of the operational side of the Conservancy's work. His first tasks were to reorganize procedures for selecting, acquiring, and managing nature reserves, and to devise ways of implementing the Protection of Birds Act of that year. The surveillance of pesticide use became his third major assignment.

The Ministry of Agriculture revised a leaflet that had already been distributed to farmers, to take account of the need to safeguard wildlife. In its Annual Report of 1956, the Conservancy commended the leaflet, and asked for reports on the effects of chemicals on wildlife where they could be substantiated with accurate data. It was to be the first of many such appeals for information.

In addition to taking all reasonable precautions, the Working Party emphasized the need for much more research and investigation, particularly on:

(1) the effects of the main sprays on wildlife on the treated land, including the effect on birds of killing large numbers of insects;
(2) the effects on parasites and predators of the insect pests against which the sprays were used;
(3) the extent to which the killing of pollinating insects adversely affected seed production in the sprayed crops;
(4) the distances over which sprays might drift, and the consequent injury to crops.

So far, the economic entomologist had provided most of the insights into the effects of pesticides on wildlife. The Working Party was impressed by the research findings of Dr A.M. Massee and Miss E. Collyer at the East Malling Research Station. Whilst the introduction of various spray-treatments had reduced the number of pests on apples from 60 to 4, they had also caused the fruit-tree red spider mite (*Metatetranychus ulmi* (Koch)) to become a major pest. By studying the life histories, host plants, and food preferences of the predators under different conditions, it had been possible for the entomologist to gain a greater understanding of the relationship between the red spider mite and its natural predators, and of how the use of chemical pesticides might have interfered with that relationship.

It was found that the use of tar-oil washes on fruit-trees in the 1920s had destroyed the mosses, lichens, and algae in which many of the mite's predators hibernated: the adoption of DNOC in the 1930s and early 1940s eliminated almost all the remaining predators of the mite. The black-kneed capsid (*Blepharidopterus angulatus* (Fall)) managed to survive, however, until DDT and gamma-BHC were incorporated in sprays from 1946 onwards. Whilst most of the red spider mites were also destroyed by the sprays, some survived and, free from all their natural predators, their numbers soon built up to pest proportions (Potter 1956). How many more cases of this type had occurred? The red spider mite highlighted the role of such research in ensuring that pesticides were used efficiently and with the minimum of side-effects.

Working parties do not in themselves solve problems. Their reports and proceedings rarely catch the public imagination. A working party does, however, establish precedents for looking at an issue. It provides opportunities for the interested parties to get to know one another. Roles are identified and working relationships forged. Although Professor Zuckerman was chairman of the Advisory Committee on Poisonous Substances for only a short time, his involvement in the subject remained strong. The concern which he, Mr Boote, and certain scientists in the Ministry of Agriculture first expressed on the Working Party was sustained, and it encouraged an ever-widening circle of people to take the subject more seriously. Wherever the forum of discussion and whatever the land use or environmental subject under discussion, there was an increasing possibility of someone being present who had 'brushed against' the pesticide/wildlife problem at some time.

The Notification Scheme

During 1955, two separate, yet related, developments occurred. The Ministry of Transport issued a Circular urging highway authorities to avoid unnecessary damage to wildlife during spraying operations, and the Zuckerman Working Party published its report. What would be the effect of these events on the Nature Conservancy, which had played a key role in preparing the Circular and was keen to 'assume responsibility for the nature conservation aspects' of pesticide regulation?

Despite the widening of the terms of reference of the Interdepartmental Advisory Committee on Poisonous Substances, there was still a strong inclination on the part of the Committee to put the wildlife aspects 'on ice'. It was widely believed impracticable to test the potential effects of pesticides on wild species. On the Committee, it was the Conservancy's task to counter this view and, in June 1956, Mr Boote drew up a memorandum describing how a wider variety of species could be included by manufacturers in their toxicity tests. Dr Massee of East Malling and Dr Potter of Rothamsted offered to give advice on breeding insect species and on the conduct of the trials.[9]

The tests usually consisted of giving groups of animals an ascending series of doses with a view to discovering the smallest dose that would kill all the individuals and the largest dose that would leave all of them alive. Because of the wide range of tolerance displayed by the individuals of any population, it was usual to measure the dose that killed half the population during the period of the experiment. Known as the LD_{50} test, the figures $LD_{50/28}$ indicated that half the population had been killed in 28 days. The test could be adapted to measure any other dose, for example LD_{10}. The graph of the percentage to survive against the size of dose usually resulted in a sigmoid, or S-shaped, curve (Fig. 3.2), although, in practice, the doses were usually expressed on a geometric scale.

The Conservancy's memorandum of 1956 also suggested that the National Agricultural Advisory Service should seek out and report any

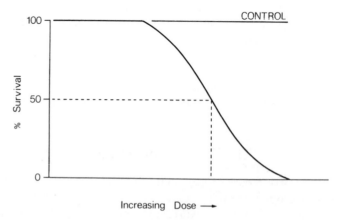

Fig. 3.2 The LD_{50} test.

damage caused to wildlife by chemicals already in commercial use. Longer-term studies should be started. The precautions advocated by the Zuckerman Working Party should also be adopted as rapidly as possible, namely that no chemical should be recommended for general use unless it was certain that the average labourer had the means and the knowledge necessary to safeguard other forms of life.[10]

Not only was the memorandum given a fair hearing when considered by the Advisory Committee, but its appearance was well timed.[11] A few months later, the Advisory Committee reached an agreement with industry for a more formal Notification Scheme, under the voluntary code of practice. Under the Scheme (later called the Pesticides Safety Precautions Scheme), the manufacturer agreed not to market a new chemical, or introduce a new formulation, or a new use for a chemical already on the market, without first providing the Ministry with details of its physical, chemical, and biological properties, its persistence, mode of action, and products into which it might break down. Before approval was given, the Ministry would consult the Advisory Committee, with its independent chairman and representatives of official departments, including the Nature Conservancy.

The Ministry believed the Advisory Committee was 'admirably suited for bringing together the expertise and viewpoints of the many Departments concerned with these problems', and for providing a useful forum for settling any differences of view. In June 1960, five independent members were added to the Committee. They were specialists in the fields of medicine, chemistry, agriculture, or wildlife. Professor A.R. Clapham, who was also a member of the Nature Conservancy committee, was nominated to represent wildlife interests.

The Committee drew heavily on the advice of a Scientific sub-committee, made up solely of scientists selected for their expert knowledge of various aspects of pesticides. The Conservancy was represented by Miss Olive Balme and, after her resignation from the Conservancy, by Dr C. Potter of Rothamsted. The sub-committee met about every six weeks. Although industry was not represented on this or the Advisory Committee, the chairman occasionally co-opted representatives to help examine specific issues.

The Scientific sub-committee scrutinized each pesticide on the basis of the data provided by the manufacturer and by drawing analogies with related compounds. It could advise the Advisory Committee that the chemical was too hazardous to use, or that only a provisional clearance should be given so that, for instance, further field tests could be held. If approval was given under the voluntary Notification Scheme, it was a condition that the labelling of the product should be in accordance with recommendations already agreed and publicized. There were also provisions for the Ministry to review the safe use of a chemical at any future time in the light of new evidence. Once approved, the product could be scrutinized under another voluntary scheme, the Agricultural Chemicals Approval Scheme, whereby it was assessed for its efficacy in controlling plant pests and diseases, weed growth, and for other crop protection purposes.

The data submitted by industry under the Notification Scheme, and the proceedings and papers of the Advisory Committee and its Scientific sub-committee, were confidential. In reply to a Parliamentary Question, the Minister of Agriculture stated that 50 notifications were received between 1957 and 1960, and a further 80 in 1961. In no case did the Advisory Committee entirely refuse clearance of a new product, but in some instances the companies did not proceed with marketing plans, 'possibly because they may have felt that acceptance of the stringent precautionary measures recommended by the Advisory Committee would have made the product insufficiently attractive commercially'. The Minister knew of no instance of products being marketed without prior clearance.[12]

The Wildlife Panel

Despite the introduction of the Notification Scheme, the Conservancy still felt that insufficient attention was being given to wildlife aspects. This was largely due to the lack of information on the effects of particular chemicals on wildlife. Nearly all reports of bird and mammal mortality related to game animals. As the regional officer for East Anglia, Dr Eric Duffey, commented, 'I wonder how often wild birds are seen dead and no-one has bothered to comment on it because the partridges and pheasants have not suffered?'. Only an extensive search explicitly for all kinds of wildlife, could settle the matter.[13]

During negotiations leading up to the Notification Scheme, the Conservancy representatives stressed to industry the benefits of being able to demonstrate that their products were harmless to bees and those forms of wildlife beneficial to farming and forestry. For their part, the manufacturers agreed in principle to submit information as to the effects of chemicals on wildlife, *if* they could be advised 'on ways and means of obtaining adequate and reliable data'.[14]

It was now up to the Nature Conservancy and voluntary conservation bodies to suggest the type of information required, and the means of obtaining it 'without placing undue strain on the co-operation of the Trade'. It was a daunting task, as became apparent when representatives of the Conservancy began to draw up a memorandum describing the routine tests that should be made before notification, and the field observations and trials that should be conducted once the products were in commercial use. There was need to assess toxicity on as many groups of birds, fish, and insects as possible, and particularly on soil organisms. All tests should embrace as many types of habitat and geographical area as possible.

The natural history and voluntary conservation bodies provided the only practical means of keeping the use of chemicals under surveillance. A suggestion from Dr Bruce Campbell that members of the British Trust for Ornithology (BTO) should help to assess the damage caused by pesticides in their respective areas was taken up with alacrity. Using drafts prepared by the BTO and the Ministry's Infestation Control Laboratory, Mr Boote compiled a draft questionnaire for use in recording any 'unusual mortality among wild birds'. There were discussions with the Laboratory as to how the involvement of the voluntary bodies might be 'dovetailed' into the surveillance work of the Ministry of Agriculture.[15] Following an informal meeting of the Ministry and voluntary bodies, Mr Boote drafted a memorandum which called for the appointment of a Wildlife Panel to assist the Advisory Committee. This would lay down criteria for pre-notification tests and trials, and for keeping pesticides under close scrutiny once they had been marketed.

The feeling that the Conservancy should take a lead on the question of toxic chemicals was strongly expressed at a meeting of the Nature Conservancy committee in November 1958. The Chairman spoke of the abundant evidence of their dangerous effects throughout the country, and a member, Colonel Charles Floyd, wrote that the problem was becoming more urgent every day. In one day, his farm manager had been pursued by no less than three salesmen from three different firms all trying to sell him new sprays for use in the following spring. The meeting called upon the Conservancy's Scientific Policy Committee (see p. 4) to 'investigate as a matter of urgency the possibility of the Conservancy undertaking large-scale research into the effect on wild life of the use of toxic chemicals in agriculture'. The attitude of the Director-General, Max Nicholson, was, however, more guarded. He stressed the 'very complicated, fast-moving nature of the problems involved, and the need for additional finance, if research work were to be undertaken'.[16]

The nub of the question was the degree to which the staff and resources of the Conservancy could be committed. There was what Nicholson (1970) was later to describe as a peculiar hitch. Although it was widely and deeply felt that the expanding use of toxic chemicals in agriculture could endanger wildlife, there was no certain proof and it was therefore impossible to obtain the necessary scientific manpower and funds to secure that proof.

Matters came to a head in early 1958, when a manufacturer accused the Advisory Committee of being unduly harsh in its attitude towards one of the firm's products. In correspondence, the firm asked whether 'the Nature Conservancy people would like to join us in looking for untoward effects on game and wildlife' during spraying trials. This was just the kind of practical collaboration the Conservancy had been striving to achieve. It could open the door for further discussions and provide an invaluable example for the Wildlife Panel. The Conservancy was, however, very conscious of the considerable responsibility that it would have to bear in participating in such trials. If the Conservancy gave its blessing to the pesticide under trial, the commercial value of that product would be greatly enhanced. Any observations or investigations would have to be made in an extremely thorough manner.[17]

The Deputy Director-General, Dr E. Barton Worthington, voiced his misgivings in a memorandum of January 1958. He wrote that this and numerous other questions would arise as manufacturers got into their stride, and farmers made greater use of chemicals. Each problem merited scientific examination. The Conservancy should limit itself to acting as an observer, or catalyst where this was desirable. Such technological questions as the formulation of standardized tests on the effects of different chemicals should be left to the Ministry of Agriculture. The Conservancy had neither the staff nor the equipment. Worthington warned that 'I think it would be unwise for us to promise much in this subject since in my view we cannot commit any more of our resources'.

The Conservancy had already abandoned further studies on the effects of spraying on roadside verges because of the manpower situation, but Robert Boote, the Conservancy's representative on the Advisory Committee (see page 28), argued strongly against a further withdrawal. There was every indication of an increase in the use of pesticides in the following spring, and there was mounting pressure from the conservation movement for closer surveillance. Boote wrote, 'it is of front rank importance in our conservation work and is a vital public relations issue'. He urged that the question should be thoroughly discussed, and the level of the Conservancy's involvement decided.[18]

Further discussion was, however, cut short by a series of economies imposed by the Government because of the general economic situation. The Director-General wrote on Boote's memorandum that increased surveillance over pesticides was one of 'the new and extended activities which it has been definitely decided by Ministers should not be started in present circumstances'. Mr Nicholson asked that the voluntary bodies be told of the decision because it was 'no use pretending we can just add more commitments without resources to cover them'.[19]

There was, therefore, no point in seeking the appointment of a Wildlife Panel: without scientists there would be no scientific data to support those representing wildlife interests on the Panel. In reply to a Parliamentary Question in July 1958, a Government spokesman confirmed that a shortage of funds and manpower had precluded any official initiative being taken on the question of toxic chemicals.[20] For the Conservancy, the reversal in expectations not only involved some loss of face, but was likely to handicap severely its representatives in their deliberations on the Advisory Committee and the Scientific sub-committee during what was clearly going to be a crucial period in the development and deployment of toxic chemicals.

Refusing to 'take no for an answer', Mr Boote prepared a paper for the Conservancy's Scientific Policy Committee, setting out the 'Effects on wildlife of the use of toxic chemicals in agriculture'. Drawing heavily on the advice given him by members of the Infestation Control Laboratory and other bodies, he explained both the current situation and the need for further initiatives to regulate pesticide use. The paper asked the Scientific Policy Committee to approve the appointment of a scientific officer to the Wildlife Panel, and to allocate scientific staff to work on toxic chemicals as the need arose and resources allowed. It emphasized that, whilst the Notification Scheme was purely voluntary, legislation would have to be contemplated 'if manufacturers showed signs of irresponsibility'.[21]

Not everyone in the Conservancy regarded the situation as being serious or urgent. The Annual Report, written towards the end of 1958, commented that there was no 'substantial evidence to prove that these chemicals are causing any important or widespread reduction in bird or mammal populations'. Boote's 'long and detailed paper' was ready in the autumn of 1958, but was placed so low on the agenda of the Scientific Policy Committee that it was not considered until January 1959. By then, popular concern was becoming more evident, and the recommendations of the paper were approved. At its meeting in March 1959, Boote informed the Advisory Committee that the Conservancy now wished to proceed with the appointment of a Wildlife Panel. The Advisory Committee agreed, and asked the Minister of Agriculture to make the necessary arrangements.[22]

It was agreed that a panel of representatives of the Ministry, Conservancy, and industry should prepare guidelines 'for industry on the type of data about wild life hazards which will be required in support of notification'. The scientific officer appointed by the Conservancy would play a key role on the Panel, discovering what was being done by the scientists working for manufacturers and sifting through the literature available to ascertain whether there were further tests that might be usefully carried out prior to notification. The scientific officer would clearly need the support of a unit devoted to toxic chemicals research.[23]

The Wildlife Panel met eventually in December 1959. It comprised two scientists nominated by the Advisory Committee, and three members each nominated by the Conservancy and industry. It was not until 1961 that a set of detailed recommendations as to what steps should be taken to protect wildlife was incorporated in a new Appendix to the Notification Scheme. This emphasized that the object was 'to suggest a few ideas which notifiers will have to expand and modify as seems appropriate'. It was impractical to lay down a rigid research programme—the scope and type of studies required would vary according to the nature of the chemicals and their proposed use. Special precautions should be taken to protect honeybees, and 'additional pertinent data (sought) . . . when the known properties of a particular chemical or its method of use, might suggest special risks to wildlife'.[24]

Between 1961 and 1973, over a hundred trials were carried out by manufacturers on the effects of new pesticides and new applications of existing pesticides, and a number of new products were refused registration on the grounds that they presented an unacceptable hazard to wildlife.

4

THE MONKS WOOD EXPERIMENTAL STATION

1958 was not a good year for the Nature Conservancy. The Director-General called it a 'pig' of a year, and hoped never to see the like of it again. The Treasury had prevented the 'already overdue expansion of staff and accommodation' and there had been 'persistent and widely publicised' criticisms of aspects of the Conservancy's work. The year ended, however, on a more hopeful note.

Two important developments kept alive the Conservancy's hopes of playing a larger part in pesticide research. Firstly, public interest in the use, or rather misuse, of toxic chemicals continued to increase. In November, more than 20 cows died on two farms in Somerset after they had grazed some weeds treated with arsenical sprays. The farmers alleged that the printed labels on the products had failed to indicate the lethal nature of the chemicals used. One of them was quoted by the *Farmers Weekly* as saying, 'without taking extraordinary precautions, which would not enter the head of the ordinary users, it is practically impossible to guarantee that human beings and livestock do not enter the sprayed area'. Both farmers called for the withdrawal of the product.[1]

Secondly, the question of whether the Conservancy should participate in research on toxic chemicals became involved with much wider discussions on the size and deployment of resources within itself. Many believed the Conservancy had to be given a much stronger sense of purpose and direction. It had to shake off, once for all, the stigma of being a dilettante organization, devoted to lobbying and bird-watching. As the results of its research began to become available, more effective ways had to be found for disseminating and applying the information. Staff would have to be found for developing the advisory, educational, and public relations aspects of the Conservancy's responsibilities. To achieve all this, there had to be a growth rate of $12\frac{1}{2}$ per cent per annum over a five-year period.[2]

The advocacy of a quinquennial phased programme assumed a more tangible form following the Government's decision to apply 'a short-term stimulus to the economy' in November 1958. Government departments were invited to put forward building and other projects that might help to increase employment opportunities. The Conservancy had been considering for some time the need for an experimental station which would act as a half-way house between the existing research stations and 'research in the field'. Whilst a new station could not be built immediately, the Government directive was taken as evidence of a general improvement in the economic climate. As Mr Boote commented, the setting up of an applied experimental station might not only provide the best location for research on the effects of toxic chemicals, but would also be well received from the public relations point of view.[3]

A centre for applied research

In a wide-ranging memorandum of January 1959, the Director-General, Max Nicholson, described how the time was ripe for the Conservancy to build a bridge between the fundamental research worker and those who used or managed land. The first need was to itemize all the different and unfulfilled requests for advice and information which the Conservancy received. Secondly, the dissemination of this advice and information should be rationalized, particularly in so far as the Conservancy, Ministry of Agriculture, and Forestry Commission were concerned. And thirdly, the Conservancy should establish 'a new and adequately staffed centre for the effective experimental study, testing, demonstration and dissemination of applied knowledge in animal and plant ecology, and the factors underlying successful management of the fauna and flora'.

The best way of achieving the third objective was to combine the following five elements, namely:

(1) an applied research and demonstration station, grouping together all applied work which could benefit by sharing common over-heads;

(2) a site well outside London forming the most convenient, practicable centre for services designed to gather and disseminate advice and information on a Great Britain basis;

(3) ready local access to plenty of land in the ownership of the Nature Conservancy suitable for trials and experiments;

(4) contact with university based scientists, and important users of information;

(5) economy in relation to the Conservancy's other commitments and functions.

The memorandum concluded by setting out the reasons why the best location for the station would be in the Huntingdon area, where communications were good and where there was the largest, and perhaps the most neglected, group of National Nature Reserves, namely Woodwalton Fen, Monks Wood, Holme Fen, and Castor Hanglands. Together they included a wide variety of arboreal, aquatic, fen and bog, heathland, and grassland habitats, both low-lying and elevated. There was unlimited potential for experimental work.[4]

Although the recommendations were approved by both the Scientific

Policy Committee, and later, the Nature Conservancy Committee, there were serious misgivings as to how far the new station might trespass on the work of other organizations. There was accordingly much interest in the more detailed proposals put forward by Mr Nicholson in April 1959. In reviewing the considerable gaps in the Conservancy's applied research programme, he cited the great embarrassment caused by the Conservancy's inability to give advice on the effects of toxic chemicals on wildlife. A new station could co-operate with manufacturers in testing chemicals or observing the effects of commercial sprays. A study could be made of those habitats where spraying was most likely to occur and damage wildlife, namely hedgerows and roadside verges. In the course of trials, the station might identify those chemicals which could be used in controlling unwanted vegetation on nature reserves and other sites of scientific significance.

A key factor was the relationship of the new station to the Infestation Control Laboratory of the Ministry of Agriculture. As early as 1954, the Conservancy had considered the feasibility of taking over some of the Laboratory's work, particularly where it related to rabbits, foxes, oyster catchers, wood pigeons, rooks, and other forms of wildlife. In February 1959, Mr Nicholson had a 'longish talk' with the Under-Secretary in the Ministry responsible for the Laboratory. He described plans for 'an applied ecological, experimental and demonstration unit in the Huntingdon area', and sounded out the possibilities of 'some kind of joint arrangement' which might extend to the actual taking-over of some of the Laboratory's staff. For its part, the Ministry stressed that the scientific work of the Laboratory was bound up closely with the advisory and executive responsibilities of the Ministry of Agriculture. Although there was clearly a risk of some overlap in research, this did not outweigh 'the convenience to the Ministry of keeping things as they are'.[5]

The Ministry had not, in fact, closed the door. In the autumn of 1958, an external Visiting Group had been appointed to investigate and advise on the role of the Infestation Control Laboratory, under the chairmanship of the Ministry's Chief Scientific Adviser, Professor H.G. Sanders. Following an invitation to meet the Group, Mr Nicholson reported to the Conservancy's Scientific Policy Committee that there was, on the scientific level, a 'complete identity between the views of the Ministry's scientific advisers, both on their staff and on the Visiting Group, and the Nature Conservancy'. As soon as the Conservancy had the resources, the Ministry should transfer the personnel and functions of the Laboratory, except where they related narrowly to pest control. The Ministry had endorsed the concept of an applied experimental station.[6]

The purpose and composition of the new station were examined in detail by members of the Scientific Policy Committee and the Conservancy's staff. They considered the question of toxic chemicals so urgent that recruitment of staff should begin before the station was built, using temporary accommodation. The need to develop vertebrate studies was also endorsed. With the approval of the full Scientific Policy Committee, the question of the new station was once again referred to the Nature Conservancy committee. All was not plain sailing. The committee minutes record that, 'after a lengthy discussion', members 'were, in general, sympathetic to the proposed setting up of the station, as outlined by the Director-General, but wished to postpone a final decision until they could be informed in more detail at a later meeting of the financial, staff and administrative implications'.[7]

At their regular meeting with the Lord President of the Council, Lord Hailsham, in July 1959, the Chairman and Director-General of the Conservancy raised the possibility of setting up an applied ecological institute where, for example, the effects of toxic chemicals on wildlife could be studied and demonstrated. They emphasized that the proposal was modest in scale and would not duplicate work done elsewhere. Lord Hailsham agreed to the proposal being studied further, but warned that the Conservancy was likely 'to face a stiff battle on costs'.[8]

The Huntingdon Working Party

The two issues of research on the effects of toxic chemicals and the establishment of a new experimental station became increasingly intertwined. The Advisory Committee had only agreed to the appointment of a Wildlife Panel on the premise that the Conservancy would provide the necessary scientific expertise (see p. 35). With this consideration very much in mind, the Nature Conservancy committee appointed a sub-committee to decide the Conservancy's involvement in this field. Dr F. Fraser Darling, Mr C.S. Elton, and Professor G.V. Jacks were appointed to the sub-committee. Valuable assistance was given by Dr C. Potter of Rothamsted, Dr I. Thomas of the Infestation Control Laboratory, and Mr F.T.K. Pentelow, the Chief Salmon and Freshwater Fisheries Officer of the Ministry of Agriculture, Fisheries and Food.[9]

At its first meeting, the sub-committee considered a paper, drafted by Mr Boote, describing how the Conservancy required three types of data. Firstly, there was the technical and chemical information provided by manufacturers under the Notification Scheme. Secondly, data were required on such ecological questions as the value of hedgerows as reservoirs for wildlife. The third set of data covered a range of ecological and chemical issues, such as how far the presence of particular chemicals might affect the ecosystem. These wider questions could only be answered by setting up carefully planned studies, including a monitoring system.[10]

The paper recommended that a Toxic Chemicals Unit should be set up in the proposed experimental station to maintain close contact with the Ministry of Agriculture and other bodies, and to investigate the ecology of the main wildlife habitats affected. It would select species suitable for use by the trade in testing pesticide products, assess the effects of chemicals on population structure, and determine the cumulative effects on wildlife. Field observations and trials would be essential, and the fullest use should be made of the voluntary bodies and their observers. The British Trust for Ornithology (BTO) and the Botanical Society of the British Isles (BSBI) should be encouraged to carry out pilot schemes in selected areas, in close co-operation with the Conservancy's Statistics Section and the Toxic Chemicals Unit, and in receipt of grants or contracts from the Conservancy.

The sub-committee recognized that the field of study would be enormous, and that the Toxic Chemicals Unit would, at least at first, be very small in size. For these reasons, the Unit should not become too deeply involved in physiological and biochemical work. There should be close collaboration with other bodies, and an efficient information service should be established to keep track of research being done both in Britain and abroad.

In its report of July 1959, the sub-committee identified three areas for particular development:

(1) to study the habitats most affected by toxic chemicals so that sound conclusions can be drawn from field experience on the effects of sprays;

(2) to study the effects of representative sprays on selected habitats in order to demonstrate the nature of the ecological changes caused by chemicals;

(3) to study the long-term effects of chemicals on selected natural populations of animals, with special emphasis on sub-lethal poisoning, and changes in reproductive and mortality rates.

The Toxic Chemicals Unit should consist of four officers of scientific grade, and four of experimental grade by 1962/63. The first three posts should be allocated to a botanist, a zoologist, and chemist.[11]

Meanwhile, the planning of the new experimental station continued. The purpose-built station was likely to cost £40 000 and temporary accommodation would be required. The most fundamental question was whether 'the MAFF contingent' would come over. In a letter to the Ministry of Agriculture in September 1959, Mr Nicholson proposed the appointment of a Working Party 'to work out exactly what research should be set in hand at Huntingdon'. In order to take full cognizance of the Ministry's experience and plans, he invited the Scientific Head of the Laboratory, Dr I. Thomas, and the Head of the Land Pests and Birds Research, Mr H.V. Thompson, to join the Working Party. The Ministry readily agreed to this. As the Permanent Secretary wrote later, 'I am sure that the closest co-operation between us will be of the greatest advantage to all concerned, and the more we can strengthen this co-operation the better I shall be pleased'.[12]

The Huntingdon Working Party reported to the Scientific Policy Committee whose chairman, Professor W.H. Pearsall, together with Mr C.S. Elton attended most meetings. The other members of the Working Party were two of the Conservancy's regional officers, Dr Eric Duffey and Dr Norman Moore, and the Principal Scientific Officer in charge of woodland research, Dr Derek Ovington. Papers were soon drafted on various aspects of the scheme. The Director-General and Pearsall, who had been Professor of Botany at University College, London, prepared a paper showing how the new station could provide advanced courses for students from University College and elsewhere. Duffey wrote a paper on the value of the nearby National Nature Reserves as research areas for conservation and other ecological studies.[13]

Encouragement came from an unexpected quarter. Following the General Election of September 1959, Lord Hailsham ceased to be Lord President of the Council, and became responsible even more explicitly for the administration of science through his appointment as Lord Privy Seal and Minister for Science (Hailsham 1975). At a meeting in October 1959, Mr Nicholson reminded him of the Huntingdon project and the favourable response from the Ministry of Agriculture. Lord Hailsham was particularly interested in the initiation of research on toxic sprays and plans for providing training courses for post-graduate students.[14]

Progress was so encouraging that, at its November meeting, the committee of the Nature Conservancy decided to ask for the approval of the Treasury in principle, so that negotiations for the purchase of a site could proceed. Mr Nicholson described to the Treasury how, 'we are all lined up to kill several birds with one stone by this very carefully thought out project'. The Ministry of Agriculture had not only approved the project but had agreed to 'their key men' being involved in the 'detailed elaboration of our plans'. Lord Hailsham was expected to approve the scheme because 'we have anticipated in an almost uncanny way some of the points on which he is asking the Research Councils to take the initiative', particularly in providing support for the universities. Many experts and interested parties had been consulted. In Nicholson's words, 'I have rarely had experience of anything affecting so many interests where all concerned were so keen and so convinced that it will give them important help in helping their problems'.[15]

Under the chairmanship of Dr Ieuan Thomas, the Huntingdon Working Party reported in March 1960, outlining the research programmes,
and estimating the needs of the four units that would make up the
Experimental Station, namely the Conservation Research Unit, the Toxic
Chemicals and Wild Life Unit, a Vertebrate Ecology Unit, and a Woodlands Research Unit. A nucleus of 37 staff should be in post by 31 March
1963 and, at a growth rate of $12\frac{1}{2}$ per cent per annum, the complement
should rise to 65. Building would be in two phases. The Working Party
adopted the recommendations of the earlier sub-committee that had
looked into the need for a Toxic Chemicals Unit. With the Vertebrate
Ecology Unit, it would need access to, and absolute control over, large
tracts of land. A minimum of 4000 yards of hedgerow would be required
for experimental purposes.[16]

The greatest bone of contention, and the main reason for the delay of
the Working Party's report, was the question as to where the new station
should be located. The Working Party visited the proposed Huntingdon
site next to the Monks Wood National Nature Reserve in January 1960,
and assessed its potential in the light of four criteria. Firstly, the station
should be situated on mixed farmland, with some woodland and well-
established hedges, the soils being of light to medium texture. Secondly,
there should be reasonable access to a variety of ecological types.
Thirdly, the site should be reasonably near a university. And, fourthly,
there should be good communications with London and elsewhere.[17]

In a progress report of January 1960, Dr Thomas recommended that
an alternative site should be found in the Cambridge area, namely on the
fen-breck margin, in a triangle between Wicken, Newmarket, and
Chippenham. In his reply, Mr Nicholson stressed the need to strike a
balance between the ideal and the practical. In his view, the only serious
objection to the Huntingdon site might be the suitability of the soils. He
was, therefore, asking Dr A.S. Watt of the Conservancy's Committee for
England, and Professor G.V. Jacks of the Scientific Policy Committee, to
give their opinion on this vital point. Unless they found 'insuperable difficulties', there was little chance of the Conservancy agreeing to an
alternative site, and, 'if they were to do so, it would mean that the whole
project would be back in the melting pot'.

When confronted with similar misgivings on the part of the Nature
Conservancy committee, the Chairman warned that 'he could not con-
template asking Ministers and the Treasury to approve an alternative
which would leave the Nene Reserves under-used, and he had reason to
expect that such a suggestion would lead to the rejection of any new
Research Station at all'. It was Government policy to concentrate
research into larger units. Very considerable importance was accordingly
attached to the report made by Dr Watt and Professor Jacks on the soil

conditions at the Huntingdon site. They visited 'the place under pretty grim (weather) conditions' in February 1960, and reported that 'the soils could be used for the general types of experiment envisaged'. In a covering letter, however, Professor Jacks reiterated some of the other objections to the site. Scientists would require access to a very wide range of biological literature, and the site was 'horribly isolated'. In presenting its report to the Scientific Policy Committee in March 1960, the Working Party insisted on submitting a separate note, once again setting out its objections to the Huntingdon site and, this time, stating a preference for a station in the vicinity of Oxford.[18]

Mr Nicholson was heard to remark on occasion, 'we must steamroller the opposition'. This was one of the times when the steamroller was busy. He emphasized once again that the Huntingdon site was, on balance, the most suitable, although it would 'certainly need more "nursing" in the early stages than one in or adjoining a fair-sized town'. Although the Treasury had been non-commital pending the submission of the Working Party report, it had allowed the Estimates for 1960–61 to include six posts, subject to the proposal for a station being agreed in principle within the year. There was clearly a risk of these posts, earmarked for research on toxic chemicals and woodland management, being lost if the Huntingdon site was rejected.[19]

In April 1960, the Scientific Policy Committee approved the report of the Huntingdon Working Party and, in the words of the minutes, recommended that the 'Director-General should be authorized to proceed with negotiations for setting up the Station as proposed'. Soon afterwards, the Nature Conservancy committee endorsed this somewhat ambiguously-worded recommendation. A day before the meeting of the Scientific Policy Committee, the Director-General had written to Lord Hailsham, asking him to approve a formal submission to the Treasury on the 'proposed experimental station near Huntingdon'.[20]

After a long discussion in May 1960, Lord Hailsham agreed to a toxic chemicals unit, and the provision of field facilities for the proposed Conservation Course at University College, London. He accepted the Conservancy's arguments that the station would be the best place for research on reserve management, and on 'the work which was needed on vertebrates in support of the Ministry of Agriculture'. Lord Hailsham was much less happy about the woodland research programme, and he insisted that the Conservancy should not become involved in farming operations. If any crops were grown in order to study the effects of spraying, these should be on land owned by the Agricultural Research Council or another body. He stressed 'the unwisdom in keeping protected populations of pest species, such as wood pigeons'.[21]

The Treasury gave its approval, in principle, in May 1960. Manchester

University and its tenant were willing to sell 66 acres of land south of the Monks Wood National Nature Reserve, but negotiations to buy a further 29 acres from an adjacent owner-occupier failed. Because this area was intended only for experimental plots, its loss did not prejudice the value of acquiring the remainder. It was hoped the farmer would change his mind later. The Treasury approved the purchase of the land available for £3,500.[22]

The Treasury had already agreed to the appointment of staff before the station had been built, and discussions followed on detailed aspects of staffing and accommodation. In view of the initially small size of the station, the Treasury challenged the need to appoint the Director at the level of Deputy Chief Scientific Officer (DCSO). Mr Nicholson drew attention to 'the diverse range of scientific subjects' to be covered by the station, and 'the inevitably delicate and complex personnel problems inherent in this kind of work'. It was particularly important that the Director should have the seniority and experience attached to the DCSO grade. In the event, the Conservancy's view prevailed and, within a few days, Dr Kenneth Mellanby was appointed. Because of that 'victory', there was no longer any chance of securing a post of Senior Principal Scientific Officer (SPSO) on the station, which the Conservancy had hoped would encourage a member of the Infestation Control Laboratory to seek promotion and appointment as head of the station's Vertebrate Ecology Section.[23]

The firm of architects that had recently designed a new laboratory block at the Alice Holt Research Station of the Forestry Commission, Messrs Cowper, Poole, and Partners, was commissioned to design the new station. In July 1960, it reported favourably on the feasibility of the project in terms of cost and timing. Because the Treasury had set a ceiling of £100 000 for building and equipping the station, only the first phase of the programme could proceed. The principal casualty was the formation of the Vertebrate Ecology Section which was, in Mr Nicholson's view, the 'only postponable part of the project'. It was a decision taken with considerable reluctance, particularly in view of the continued hope that some of the personnel of the Infestation Control Laboratory would join the section.[24]

The architects chose a layout and design for the station which would permit simple and economical adaptation as staff grew and changes occurred in the research programme. Because of the character and beauty of the site, and recent criticisms of new buildings in the countryside, great care was taken to integrate the buildings into the landscape (Mellanby 1963). The provision of electricity to the site was no problem, but it proved impossible to provide mains gas. The cost would have meant the sacrifice of four laboratories. The most serious problem arose over the supply of water and, for a few weeks in the summer of 1960, this threatened to jeopardize the entire project. The existing public supply was already inadequate, and there was no chance of securing an adequate supply of good-quality water from a well drilled on the site. It was with profound relief that the Conservancy learned in August 1960 that the Rural District Council could make available a supply of adequate pressure from Abbots Ripton.

At its November meeting in 1960, the Nature Conservancy committee considered the name to be given to the new station. Lord Hurcomb preferred the 'Alconbury Experimental Station', but Sir Joseph Hutchinson objected that 'Alconbury' meant only two things in that part of the East Midlands, namely Americans and atomic bombs! Before the committee's next meeting, it was discovered that the site was not even in the parish of Alconbury and, after some further debate, the name 'Monks Wood Experimental Station', was accepted. By adopting the name of the National Nature Reserve, it would be possible to avoid having to insert two names into one small area of the map![25]

The Toxic Chemicals and Wild Life Section

Right from the start, considerable thought was given to who should lead and make up the research units on the new experimental station. In correspondence, Mr C.S. Elton warned that 'if you are going to cannibalize the Conservation and Fundamental Research branches of the Conservancy to set up the Huntingdon Station, it will fulfil the worst fears of people like Diver, and some of mine'. Research programmes would be sabotaged if scientists were moved around every few years—as happened in the administrative sections of the Civil Service.

Clearly, this was a danger to be avoided, but Mr Nicholson warned that the morale of the Conservancy would suffer if the station were opened with entirely 'new men'. The new station 'should contain a good leavening of people who really know what we are at and have the confidence and full understanding of their colleagues'. In the event, Dr Norman Moore was appointed head of what became the Toxic Chemicals and Wild Life Section, and Dr Eric Duffey of what was to be the Conservation Research Section. Both had been regional officers, and had 'kept up their research interests'. They were well qualified to help provide that 'much needed bridge between conservation work and research'. Dr J.D. Ovington, who had already developed the woodland research at the Merlewood station, also joined Monks Wood.[26]

A Vertebrate Research Unit was never formed. Hopes that some of the responsibilities and personnel of the Infestation Control Laboratory might be transferred to the Conservancy were never realized. The Ministry of Agriculture concluded that the research and advisory functions of the Laboratory were so integrated that it would prove difficult to transfer the former and leave the latter.[27] The Laboratory moved into new accommodation at Worplesdon. The potential benefits to the Conservancy of such a transfer of resources and expertise passed into the 'ifs and buts' of history. The Conservancy lacked the posts to create a Vertebrate Ecology Unit unaided, and the Toxic Chemicals and Wild Life Section lost not only the in-house support it might have expected from such a Unit, but also a potential ally in pressing for a large experimental area for conducting field trials.

Although the Toxic Chemicals and Wild Life Section was formally set up under Dr Moore in January 1960, some time would elapse before it was a functioning unit. The Conservancy had only decided to establish the Section in January 1959, and the Treasury had only given its approval, in principle, in November. It was not until mid-December that Dr Moore's appointment was confirmed, and he forecast that almost all of 1960 would be taken up with handing over his responsibilities as regional officer, writing up his backlog of scientific papers on the buzzard, dragonflies, and heathland conservation, and in helping to plan the new experimental station. Temporary accommodation would be needed and staff had to be recruited. At the most, Dr Moore doubted whether the Section would 'really exist as a properly functioning research body, capable of giving advice, until well into 1962', and advice based on its own research could not be forthcoming until several years after that.[28]

Dr Moore left Dorset in August 1960 and, for the following four months, his home in Swavesey was the Section's base. Secretarial and library facilities were kindly made available by Dr Alistair Worden at Cromwell House, in Huntingdon. The Conservancy eventually succeeded in renting temporary accommodation for the Section in St Ives. The Treasury authorized three posts for the Section in the spring of 1960, one of which, the post of Principal Scientific Officer, was already filled by Dr Moore. A soil zoologist, Dr B.N.K. Davis, was appointed as a Scientific Officer in September 1960. When an agricultural chemist, Mr C.H. Walker, joined the Section in January 1962, the University School of Agriculture offered laboratory facilities in Cambridge.

The first priority was to draw up a more detailed programme of research for the Section, and to make contact with the many other organizations that were somehow involved in the wildlife/pesticides issue. Toxic chemicals were likely to affect wildlife in two ways, namely, through the introduction of new mortality factors and by a reduction in the areas of habitat available to wildlife. In a paper to the Pesticides Group of the Society of the Chemical Industry in March 1962, Dr Moore

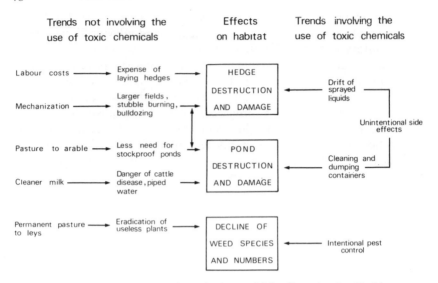

Fig. 4.1 Some parallel trends in agriculture which affected animal habitats.

emphasized that no species lived in isolation. It was part of 'an enormously complex society of interacting elements—an ecosystem'. In order to understand the ecological effects of toxic chemicals, all the other constraints imposed on a species had to be taken into account (Fig. 4.1). The distribution and abundance of species were much influenced by food supply and the suitability of conditions for reproduction. Whilst the area of habitat available was important, the size of population usually reflected the balance struck between the ability of a species to increase and such causes of death as disease, predation, and accident (Moore 1962a, 1962b).

In deciding which aspects of pesticide research to focus on, Dr Moore rejected the conventional wisdom of concentrating on selective weedkillers or of trying to establish the toxicity of different compounds to wildlife. Beyond discouraging the use of such weedkillers on sites important for wildlife, and stressing the need to minimize the risk of spray drift and other side-effects, there was very little the Conservancy could do in practice because weedkillers were so valuable to farming. Dr Moore recommended that scientific attention should be devoted much more to understanding the effects of insecticides on wildlife. Although used on a smaller scale, they were potentially much more harmful to wildlife because of their persistence in the environment.[29]

The Section should focus on the longer term effects of pesticides. Not only were these likely to be more significant ecologically, but no other research body was investigating them in any detail. So far, the emphasis in research had been on laboratory tests and *ad hoc* observations. Whilst these had proved that particular compounds killed many individuals of a species under given conditions, the more fundamental, ecological questions remained unanswered. Was the concurrent decline in many species due to toxic chemicals or some other, coincidental factor? To what extent could populations of different organisms adapt to different sprays? What were the combined effects of the applications of different compounds? What were the long term effects of sub-lethal poisoning on individual species? These questions could only be answered through ecological and long-term research, conducted in the field under experimental conditions.[30]

The formative years of the Section were not wasted. Not only did the Conservancy strike out on a distinctive and challenging course, but one that proved both relevant and realistic. By concentrating on the longer-term implications of using insecticides, the Section happened upon what proved to be the most dangerous side-effects of their use. It was well placed to perceive the full implications of what was happening in the environment. Such was the existing state of knowledge that it was possible to make a major contribution to pesticide/wildlife research with relatively modest resources.

All this was not apparent immediately. Whilst the Conservancy's Scientific Policy Committee soon approved Dr Moore's proposed programme, it proved much harder to convince others and thereby secure the patronage on which the growth of the programme depended. Paradoxically, by the time the validity of the Section's approach was established, it seemed so obviously correct that its success was taken for granted. The uphill struggle to have the programme accepted was either overlooked or forgotten.

The (Sanders) Research Study Group

Throughout these deliberations on the size, shape, and functions of the Toxic Chemicals and Wild Life Section, important changes were taking place, both in regard to the perception of the pesticide threat and to the emergence of an adequate response to that threat. Both were to have a profound influence on the pattern of development within the Toxic Chemicals and Wild Life Section.

By the autumn of 1959, serious misgivings were being voiced in parliament. Viscount Elibank put down a Parliamentary Question, drawing attention to the losses of bird, insect, and mammal life caused by 'certain all-obliterating toxic crop sprays'. The Nature Conservancy warned the Ministry of Agriculture against giving a 'complacent and even rashly reassuring' reply. There was a real danger of any assurances being proved wrong. In his reply in the House, the Joint Parliamentary Secretary of the Ministry attributed the high mortality of bees to the exceptional weather which had necessitated spraying at precisely the same time as the weeds, visited by the bees, were in flower. This and other problems were being kept under close surveillance by the Advisory Committee on Poisonous Substances. The minister emphasized that the risks incurred by spraying had to be seen in the context of the great benefits otherwise conferred on agriculture by the chemicals.[31]

In the House of Commons, John Farr used the occasion of his maiden speech to give notice of a motion calling on the Government to set up a commission of enquiry into the use of chemical sprays. Not only was little known about the short-term effects of spraying, but practically nothing about the longer-term repercussions. He called for 'a permanent research centre whereby all problems connected with toxic sprays and their application can be investigated'.[32] Notice of the Ajournment Debate brought to a head proposals that the Ministry of Agriculture had been considering for the appointment of a Research Study Group to investigate the question of toxic chemicals. In a letter to the Conservancy, a spokesman wrote that the Minister was very concerned over the whole question, but he had decided that there was no case for 'compulsory legislation' as long as the voluntary Notification Scheme worked well—at least in dealing with day to day problems. The greatest weakness was the lack of information about the longer-term effects of the chemicals on consumers, farm crops, and stock, as well as on wildlife.[33]

In the closing minutes of the Adjournment Debate, the Joint Parliamentary Secretary announced that, in consultation with the Minister for Science and the Minister of Health, the Minister of Agriculture would appoint a small group of scientists 'to review the situation and to make proposals as to whether more research work generally is necessary'. The Parliamentary Secretary continued:

Our declared aim is to maintain our present efforts to safeguard all interests; to refuse to be rushed into action which could not be justified by experience or logic; to encourage by all means in our power the establishment of an even firmer and broader foundation of our knowledge on these matters; and to continue to do all we can prevent hazards from the use of toxic chemicals in agriculture.

The announcement of any committee of enquiry leads to fears lest it may be another device on the part of government to put off the day of decision, and such fears were implicit in an article published in the *Observer*, written by Lord Hurcomb, the President of the Society for the Promotion of Nature Reserves and a member of the Nature Conservancy committee. Whilst it was important, he said, to review research needs, it was even more essential that steps be taken to prevent damage to wildlife caused by carelessness, thoughtlessness, and the generally excessive use of pesticides. Whilst a review of long-term research might help to reconcile methods of greater food production with nature conservation, it could add nothing to the more immediate and effective enforcement of precautions already known to be available.[34]

The Research Study Group met under the chairmanship of Professor H.G. Saunders, the Chief Scientific Adviser to the Minister of Agriculture. The other 11 members were leading scientists in their respective fields, and seven were members of the Advisory Committee or its Scientific sub-committee. They included Professor A.R. Clapham, who recalled later that most members were concerned initially with only the effects of chemicals on human beings and their food. The anxieties of the naturalist were dismissed as 'mere sentimentality'. The Group's attitude began to change, however, as reports of bird mortality increased, and the press and parliament called for remedial action.[35] In its report of September 1961, the Research Study Group described how members had become impressed by the depth of feeling aroused, and by the difficulties encountered in seeking to resolve some of the threats to wildlife posed by the use of pesticides (Research Study Group 1961).

The first task of the Group was to define the role of research in helping to resolve these threats. The Group's report stressed that research alone could not decide whether 'the economic advantages accruing from a given method of pest control outweighed from the collective as well as the individual standpoint, the various possible disadvantages of resulting changes in population of wild plants and animals'. These decisions were essentially political in nature, but the Group was convinced of the need for any administrative decision to be based 'on factual information obtainable only by research of one kind or another'. The Group deprecated the fact that some of the evidence had been based more on apprehension of what might happen rather than 'on direct observation or proven facts'. Very few suggestions were made to the Group as to specific lines of research that might be pursued.

Most of the suggestions that had been made were contained in a memorandum drawn up by Dr Norman Moore. The Conservancy was under no illusions as to the importance of the Research Study Group for the future of the toxic chemicals research programme. In the words of Dr Worthington, the Group's findings were 'likely to form the shape of things to come', and great care was accordingly taken in preparing the Conservancy's submission.[36] Dr Moore described how the Conservancy accepted the use of chemical sprays as an integral part of modern farming, and how it believed wildlife conservation could be made compatible with good farming if spraying practices were based on an adequate knowledge of:

(1) the factors controlling wildlife population on farmland;
(2) the physical behaviour of sprays (drift, persistence, etc.);
(3) the biological effects of sprays on plants and animals

Research was need on all these aspects. In its report, the Research Study Group followed Dr Moore's example by drawing a distinction between research needed to make short-term measures more effective, and the initiation of longer-term studies.

With regard to short-term research, Dr Moore believed that the two principal needs were the setting up of a post-mortem service as a means of establishing the cause of death of birds found in the field and any possible effects from toxic chemicals. The Research Study Group .readily approved of this proposal, but was less happy about the second suggestion, namely, to carry out the same toxicity tests as used by industry on a much wider range of plants and animals. By comparing the results of these wider studies with those made on the indicator species used by industry, Moore argued that it should be possible to appraise the value of these indicators. The Group doubted whether a reliable extrapolation could be made in this way.

Turning to longer-term research, the Research Study Group approved a proposal for a nation-wide monitoring system, whereby any rapid decline in population could be detected, and warnings given as to any dangerous consequences. The Group also supported a proposal for a survey of agricultural habitats in order to relate the changes brought about by spraying to the other factors causing deaths among wildlife. It was important to know how the minor habitats of the farming environment, such as the covert, ditch, and duck pond, were effected by spraying. These marginal areas were the principal reservoirs for wildlife on agricultural land.

In its submission, the Conservancy identified the general fields of responsibility in research. The collection of field data for the post-mortem and monitoring services would need numerous and widespread observers, and was best left to the Pest Officers of the Ministry of Agriculture, working in conjunction with the amateur naturalists' organisations. The analytical side of the post-mortem service would need extensive laboratory equipment and staff, and was again best left to the Government Chemist or the Ministry of Agriculture. The Conservancy should concentrate on those aspects requiring an ecological approach and longer-term experimental work, for which the continuity of land tenure and services would be essential.[37]

On the assumption that six scientists would have been allocated to the Toxic Chemicals and Wild Life Section by 1963, the Conservancy outlined its tentative research programme to the Study Group. After his oral examination, lasting an hour, Dr Moore confessed that some members found the programme too ambitious. They had serious reservations about the practicality of conducting experiments on the long-term effects of continuous spraying, including studies on the effects of sub-lethal dosages and genetic change. Dr Moore concluded, 'we will doubtless run into a number of blind alleys, but there should also be breakthroughs. I think we should take the obvious risks and tackle the long-term ecological problems'.

In its report, the Research Study Group found that no important aspect of research on the use of toxic chemicals was entirely neglected, but the scale of effort was, in some cases, not commensurate with the need. The Group recommended that fundamental research should continue into the ways in which pesticides worked, as a basis for finding less toxic and more selective chemicals. There was a need for more research on the breeding of varieties of crops more resistant to diseases and pests, on the development of pesticidal resistance within pest species, on the toxicity of pesticides to mammals, and on the value of repellents. There should be further studies of the interdependence of wildlife species, including population studies, in order to help predict the effects of new compounds.

Turning to those aspects of research which should be intensified, the Group recommended an examination of hazards to operators using the chemicals, and surveys of residue levels in human food and of the effects of pesticides on wildlife, with special attention being given to the protection of bees and other beneficial insects. It was important to improve methods of analysing pesticide residues. Not only were post-mortem diagnoses of large samples required, but the standards had to be set for determining the level of chemicals required to cause death. This would help prevent the wrong inference being drawn when traces of chemicals were found.

The Group was particularly concerned that investigations should be carried out on the build-up of pesticides in the soil, on the intake of residues by worms, slugs, snails, and other invertebrates in or above the soil, and the effects on any birds or mammals that ate them. Special attention should also be given to the contamination of bodies of water, whether directly through the use of herbicides on water-weeds or indirectly through surface run-off.

Turning to the machinery for regulating the use of pesticides and for initiating research, the Research Study Group was encouraged by the fact that, despite the increasing use of pesticides, the proportion of land treated with the more toxic compounds had declined by a quarter. Schradan had been almost entirely replaced by less toxic compounds, and

DNOC by the equally efficient, but less dangerous, herbicides. Each year, cattle had died from eating potato haulm sprayed with lead arsenite to kill the haulm. The accidental contamination of a static water supply, during spraying operations, had led to six persons being poisoned, one of whom died. In October 1959, an agreement was reached by the Inter-departmental Advisory Committee whereby it was decided to withdraw lead arsenites from sale by the end of 1960.[38]

This was the first instance of a voluntary ban being imposed. A circular from the Minister of Agriculture to the chairmen of the county agriculture executive committees emphasized how great a sacrifice was being made. The manufacturers had agreed to stop making a product for which there was a ready and regular market, and farmers were foregoing an effective aid to husbandry in the wider public interest. About an eighth of the potato crop had been treated with arsenites.[39]

The Conservancy's representative on the Advisory Committee, Mr R.E. Boote, agreed that the ban marked an important stage in the voluntary Notification Scheme. If the industry collectively accepted the recommendation, the Scheme would be considerably strengthened. On the Advisory Committee, the Conservancy was encouraged to press even more strongly for the retention of only those pesticides which were clearly better than all others, or which involved comparatively little risk to wildlife. Mr Boote reported that there was increasing support for this view, and at least one product was withdrawn in 1961 because the manufacturer could not prove the product was both necessary and harmless.[40]

In the light of the various assurances given to the Research Study Group that the use of chemicals was being closely scrutinized and regulated, the Group rejected proposals that the voluntary procedures should be replaced by a statutory one. The present procedures were flexible, and provided opportunities for joint consultation. During the negotiations leading up to the establishment of the Notification Scheme, the industry had agreed to supply data on the toxicity of products to wildlife, and the value of this collaboration had been testified to, for example, at a meeting of natural history and nature conservation bodies convened by the Conservancy in September 1957. Further encouragement was derived from an invitation extended to the Chairman of the Conservancy to attend the annual dinner of the Association of British Manufacturers of Agricultural Chemicals (ABMAC) in 1960. Representing virtually all British producers of agricultural chemicals, the Association sought to promote efficiency and safety in the use of the industry's products. In his speech at the dinner, the Association's President spoke of their desire to establish 'a real working partnership'.

The Research Study Group believed that the Advisory Committee and its Scientific sub-committee were well placed to identify problems at an early stage and to encourage a co-ordinated and effective response. It was better to encourage the existing research bodies to develop their potential, than to create new ones. If their staff were expanded, these bodies would be capable of carrying out the necessary field surveys and residue analyses. Fundamental and applied research were properly the responsibility of the research councils, and the Group welcomed the establishment of a new research station within the Nature Conservancy, partly devoted to studying the effects of pesticides on wildlife. Amateur naturalists and their societies should become involved in the survey work wherever possible.

In the Conservancy's view, the findings of the Research Study Group fully vindicated the decision to create the Toxic Chemicals and Wild Life Section. An agricultural chemist, Colin H. Walker, was to join the Section in January 1962, and a botanist, J. Michael Way, in the following April. With the appointment of an Assistant Experimental Officer, an information service was provided. This was of particular value in view of the fact that about 300 relevant papers were published each month in a wide range of periodicals, many of them obscure. The Section could not have functioned adequately without the service, and the role of the Conservancy on the Scientific sub-committee of the Advisory Committee, and on the IUCN Committee on the Ecological Effects of Chemical Controls, would have been seriously diminished.[41]

Privately, the Nature Conservancy found the report of the (Sanders) Research Study Group disappointing. Far from persuading ministers and the Treasury of the urgency of the problem and the need for much greater resources to tackle it, the report gave the impression that all would be well if only a little extra money were spent on existing research programmes. In a memorandum of October 1961, Dr Moore warned that the Section would soon become seriously handicapped by a lack of essential resources. The most conspicuous of these was the continued absence of any facilities for chemical analyses, particularly those for measuring residues found in corpses and in the soil. Not only were they essential for both short- and longer-term studies, but their absence was particularly frustrating in view of the range and accuracy of analyses that had recently been perfected. If no other organization could be persuaded to take on the task, the Conservancy would have to instal the post-mortem service.[42]

The most obvious body to provide the service was the Laboratory of the Government Chemist, with its unassailable status as an independent arbiter in matters of dispute. The Laboratory had played a conspicuous part in revolutionizing the trace-analysis of organochlorine pesticides. The introduction of gas-liquid chromatography with electron-capture detection had greatly increased the sensitivity and, to some extent, the selectivity of analyses. A Pesticides sub-division had already been formed, primarily to assist the Inter-departmental Advisory Committee in assessing the presence and significance of residues in human food-stuffs (Laboratory of the Government Chemist 1962, 1974).

In February 1962, a meeting was held to discuss the feasibility of the Laboratory acting on behalf of the Conservancy. Fears were expressed lest this might lead to similar requests from other bodies, thereby making further demands on staff and resources. In the course of discussion, Mr Boote, for the Conservancy, drew attention to the importance which the Zuckerman Working Party had attached to concentrating all analytical work on one site. The Treasury was much more likely to agree to additional staff being appointed to the Laboratory than to the creation of a new and separate analytical section in the Conservancy. The discussions led to an agreement whereby the laboratory carried out analyses on a repayment basis. For the Conservancy, the agreement marked the beginning of an important phase in its toxicological work.[43]

There was plenty of scope for despondency in another field of activity, namely the setting up of a monitoring system that would be sufficiently sensitive to detect changes in wildlife populations before they became generally obvious. The returns from the first year of the Common Bird Survey, organized by the British Trust for Ornithology (BTO), were few, patchy, and inaccurate, and when Dr Moore suggested that various entomological societies might record the presence of 67 species, or groups of species, in seven types of habitat, he was told that this was far too ambitious.[44] It seemed that little support could be expected from the voluntary bodies for some years—a conclusion that turned out to be too pessimistic. Once members were motivated, their contribution was to be outstanding.

Perhaps the most enduring disappointment proved to be the failure of the Section to secure adequate facilites for field experiments. Whilst landowners were prepared to give the Section opportunities for observation, none would agree to the Section having long-term control over spraying and cropping. In a paper of May 1961, Dr Moore recommended that the Conservancy should own or lease a tract of farmland, or at least have an agreement with the owner similar to that of a Nature Reserve Agreement. A considerable sum of money might be needed to compensate the owner for lost production. Whichever method was used, a farm manager with agricultural scientific training would be needed to supervise the farm work in order to ensure the right conditions for experimental work.

The proposal was rejected by the Chairman and Director-General. By adopting such a course, the Conservancy would lay itself open to further charges of being profligate in its demands for land. Many years later, the

Earl of Bessborough, who was Lord Hailsham's Parliamentary Secretary, recalled how Lord Hailsham, as Minister for Science, had 'made it plain that the Nature Conservancy was trying to obtain too much land' for nature reserves and other purposes (Select Committee 1980). The Toxic Chemicals and Wild Life Section was recommended to approach the Ministry of Agriculture, the other research councils, or universities, in order to discover whether the facilites of an experimental farm could be shared. It was, in any case, better to have a scatter of experimental sites, rather than a single site, when attempting to discover the range of changes that might be brought about by the use of toxic chemicals.[45]

The ambivalent attitude of the (Sanders) Research Study Group, and the delays and setbacks in securing the pre-requisites for research, highlighted the difficulties experienced by the Toxic Chemicals and Wild Life Section in seeking to pursue a research programme that was both challenging and consistent. As the attempts to secure a post-mortem service indicated, the approval of a research programme was not, in itself, a guarantee of a co-ordinated response, even within the 'government service'. It took time for the ramifications of any decision to be worked out, and for each party to be given its alloted task. Whilst politicians might call for the vigorous pursuit of some line of research, there were other political factors to be taken into account. In considering whether an experimental farm would be required, the Conservancy could not run the risk of being criticized for 'sterilising' land and taking over sectors of agricultural research. Even when the significance of the Section's research programme was at last recognized and additional resources were provided, the problems were not over. The more important the research, the more people wanted to influence the shape and direction of that research.

These wider considerations imposed a severe constraint on the implementation of research programmes and on the day to day conduct of research. Somehow, through it all, the staff at the new experimental station had to conduct their experiments, and publish papers.

5

SEED–DRESSINGS

During the 1950s, the damage inflicted on wildlife by pesticides was generally dismissed as inevitable and localized, without any major or lasting effect on species populations. Not only had there been no major incidents but there was a trend towards less toxic compounds being used. The regulations on pesticide use had been improved and there was now better training for operatives (Edson 1964).

Not everyone believed things were getting better. The misgivings of the Conservancy were so strong that they led to the formation of the Toxic Chemicals and Wild Life Section at the new experimental station at Monks Wood. Whilst the station was being planned and built, and the (Sanders) Research Study Group was deliberating, a major crisis developed over a new range of compounds used for dressing seed-corn. The ramifications were to have a profound effect on the development of the Section, and the significance attached to its research.

For many years, fungicidal seed-dressings had played a major part in reducing losses caused by a number of seed-borne organisms. In order to prevent a serious build-up of diseases, the seed was treated annually or periodically with a fungicide. From 1947–48 onwards, the effectiveness of these dressings was greatly enhanced by incorporating insecticides. The dual-purpose dressings could protect crops not only against seed-borne diseases but also against such invertebrate pests as the wheat bulb fly and wireworm.

The wheat bulb fly (*Loptohylemyia (Delia) coarctata* Fall.) was probably the most serious insect pest of winter wheat, particularly in the eastern counties of England (Fig. 5.1) (Gough 1957). The adult flies laid their eggs on the bare soil in late summer and by early October it was possible for the Ministry of Agriculture to warn farmers of the likely level of attack in the following spring, based on egg-counts. The larvae emerged from mid-January onwards, and entered the plant below ground at the first node. Although the sowing of seed early in the season and the use of fertilizer enabled the plant to be well developed by the time the larvae emerged, the only certain method of controlling damage was to dress the seed with an insecticide. Increasing numbers of farmers bought their seed ready-dressed, thereby avoiding the need to dress the grain themselves or resort to spraying later in the season. They were advised to sow the treated seed at a shallow depth so that the parts of the plant most vulnerable to attack, namely those between the seed and the surface, would be well within 'the zone of protection' of the insecticide. This

58

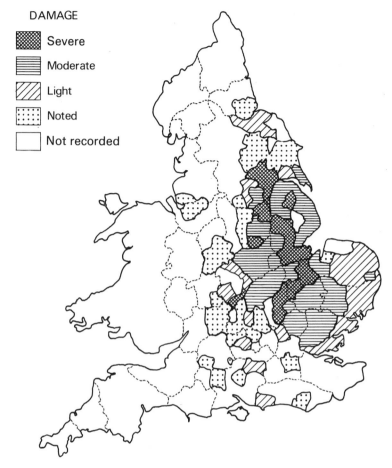

DAMAGE

- Severe
- Moderate
- Light
- Noted
- Not recorded

Fig. 5.1 Damage caused by wheat bulb fly, as recorded by the National Agricultural Advisory Service in 1953.

made it even easier than before for birds to find and ingest some of the treated corn.

The first insecticide to be used in dual-purpose dressings was benzene hexachloride (BHC). Unfortunately, it was found to be mildly phytotoxic, particularly in wet seasons, which meant that when the dressed corn was tested under the requirements of the Seeds Act, there appeared abnormalities in seed growth of up to 80 per cent. Whilst this did not affect the ultimate yield of the crop, the abnormalities severely reduced the value of the dressings under the Act. Particularly after the very wet season of 1955, farmers were encouraged to make greater use of a new range of chemicals, aldrin, dieldrin, and heptachlor, which did not behave in this manner. They were also much more persistent as dressings on winter wheat, thereby obviating further spraying against wheat bulb fly in spring.

Aldrin and dieldrin became available in Britain for experimental purposes in 1952/53 and field trials on a commercial scale began in 1953/54. The fact that they were toxic to birds and animals did not arouse the concern of manufacturers and agriculturalists because the seed was intended to be buried in the ground. As Mellanby (1964) remarked later, no one foresaw that some of the seed was bound to be left on the surface, especially in poor sowing conditions, and that birds were likely to dig up some of the seed and consume it.

The first deaths that could conceivably be linked with these chemicals were reported in March and April 1956. Wood pigeons (*Columba palumbus*) were the most affected, with heavy mortality among pheasants in some parts. There were reports of pigeons dropping dead in mid-flight and from their roosts in trees. By the end of April, 200 corpses had been sent to the Game Research Station and many more to the Veterinary Laboratory of the Ministry of Agriculture. As A.D. Middleton wrote, no one was able 'to determine exactly which chemical—if it is one of the seed dressings—is responsible'. There was, however, 'a good deal of circumstantial evidence indicating that a lot of the pigeon mortality is due to the new Shell "Dieldrex".[1]

Suspicions were further aroused by a full-page advertisement placed by Shell Chemicals Ltd in *Sport and County* on 18 April 1956. Unfortunately, the Agricultural Division of Shell Chemicals Ltd refused to give copyright consent for it to be reproduced in this book. The advertisement showed wood-pigeons standing among cabbages and beet. Under the heading 'Pigeons on your mind?' it described how evidence was accumulating that suggested that pigeons were allergic to Dieldrex 15, the dieldrin-based insecticide for 'fly' and other insect pests. This, Shell commented, seemed a fascinating possibility, and a number of enterprising farmers had already decided to try out Dieldrex against pigeons: the birds caused considerable damage to crops, and the pesticide was cheap. The advertisement concluded with an invitation to 'anti pigeon pioneers' to send in progress reports.

This advertisement illustrated beyond doubt that the company knew that certain uses of dieldrin could bring the compound into contact with birdlife. The Game Research Station stressed that 'suspicions, supported by some circumstantial field evidence' had fallen upon the seed-dressing, Dieldrex A, which was intended for use primarily against wireworm. Dieldrex 15 was not being promoted as a seed-dressing. It nevertheless contained the same active principle, which was known to be 'more toxic than most of the other insecticides used'. The British Field Sports Society protested that any chemical which harmed pigeons would have a similar effect on game and other wild birds. The Nature Conservancy found the advertisement 'shattering'. The Ministry's Infestation Control Laboratory agreed with the Conservancy that such an advertisement could 'well lead to the use of insecticides for purposes for which they were not intended and lead to widespread poisoning of birds'. The use of poisonous or stupefying baits without a licence (except in very special circumstances) was an offence under the Protection of Birds Act of 1954. The advertisement was not seen again.

Despite this outcry, Mr Middleton emphasized that there was no method of post-mortem analysis to prove that any of the seed-dressings could cause high bird mortality. The position changed dramatically, however, when, in June 1956, the Infestation Control Laboratory informed the Conservancy that a series of preliminary analyses indicated that dieldrin was toxic to pigeons and probably to other birds.[2] In a paper published later, Carnaghan and Blaxland (1957) described how the Veterinary Laboratory had soon discounted virus or bacterial diseases as the cause of the high bird mortality in some areas. In view of the time of year and state of the carcases and visceral organs, feeding trials were devised in the Laboratory to discover whether the birds had been poisoned. Although grain dressings of organo-mercurial and gamma-BHC were found to be non-toxic, those pheasants and pigeons fed on dieldrin-dressed grain died within 4 to 10 days. The birds became hunched up and listless about 48 hours before death, and their flight became ungainly. Severe convulsions preceded death. Although this period of illness appeared to conflict with observations in the field of birds dropping dead out of the sky, the fact that the digestive tract of many of these field specimens was empty suggested that they too had been suffering a more protracted illness than had at first seemed apparent. On the basis of the circumstantial evidence and the laboratory tests, the paper in the *Veterinary Record* concluded that dieldrin dressings had caused the exceptionally heavy mortality. The birds would have been weakened by the severe late weather, and the dry summer might have caused the dressings to adhere to the grain for a longer period than normal. It was significant that mortality ceased quite abruptly once the corn had sprouted.

The Ministry's Pest Officers were asked to keep fields sown with dressed seed under surveillance in 1957 and, by early April, reports of further deaths were received. In 8 out of 10 cases, they coincided with the main sowing period and, of the small proportion of corpses picked up, 238 out of 240 were grain-eating species. About 160 were wood-pigeons, 33 were pheasants, and 30 were rooks. The manufacturers conceded that dieldrin dressings were likely to be more toxic to birds than the other dressings in common use, but there was a fair body of evidence to suggest that birds were generally repelled by them. They claimed that the total number of birds affected was small, and that the problem was therefore exaggerated, and that by far the largest number of birds to suffer were wood-pigeons, which were unmitigated pests to farmers. In any case, the manufacturers argued, the loss of a few birds was a small price to pay for the benefits to be derived from the new dressings.[3]

The Inter-departmental Advisory Committee on Poisonous Substances appointed a Panel of the Scientific sub-committee in 1957 to investigate the evidence and make recommendations. It not only studied the toxicological evidence available, but issued a circular to the National Agricultural Advisory Service and to Pest Officers, as well as to the British Trust for Ornithology and British Naturalists' Association, asking for details of unusual bird deaths. Only 30 replies were received. The Panel concluded that, in view of the large acreage sown with dieldrin-dressed seed, and the number of people who were asked to look out for evidence, the number of incidents suggested that the use of dressings was not a major threat to birdlife. As the birds most seriously affected were wood pigeons and other seed-eating species, the losses were likely to be beneficial to farming. Although some gamebirds were affected, there was not enough evidence to assess what proportion of the population was affected.[4]

Turning to the future, the Panel saw little point in encouraging and relying on the types of reports that had previously made up the sum total of the evidence available. Such reports gave no precise details as to the distance between the potential source of poison and the scene of death, or the time that had elapsed between consumption and death. None of the reports provided a satisfactory basis for assessing what proportion of the total natural population of each species has been adversely affected. On the other hand, the Panel recognized that there was considerable public unease over the poisoning of birds, and that it was illegal to expose dressed grain with the express intention of killing birds. It advised that one or more persons should be employed to make detailed observations over defined areas. There should be further studies made of the fact that, under experimental conditions, some birds appeared to reject dressed, as opposed to undressed, grain. This suggested that all seed-corn might be treated with repellents as a means of reducing bird damage and damage to birds. The Panel was disbanded in early 1959 and neither proposal was adopted.

Hard on the heels of every assurance that the problem was resolved or insignificant came further reports of exceptional mortality. In June 1959, the Conservancy received a letter from a Lincolnshire farmer describing how dozens of wood-pigeons, pheasants, and small birds were found dead after the spring corn was sown. Two dogs and cats had died within a day of eating the corpses of some of the affected birds. Another unsolicited letter came from the head-keeper of a large estate near Scunthorpe. He wrote that he had complained to farmers about poisoning for over nine years. On one occasion, he had picked up '104 Greenfinches and Chaffinches, 16 Pheasants, 7 Partridges and roughly over 200 Wood-Pigeons'. In his words, 'they had been killed by sowing Autumn Wheat. Many of my keeper Friends have told me that they have experienced the same thing'.[5]

The scale and frequency of these reports helped to refute suggestions that these deaths were the products of freak combinations of circumstances. Whilst such letters suffered from all the limitations identified by the disbanded Panel, they were writtten by men with an extensive and intimate knowledge of their respective parts of the countryside. In acknowledging the letters, the Conservancy expressed its considerable concern and asked for more details. The Conservancy was particularly worried about references to birds of prey having died after eating 'poisoned birds'. The British Falconers' Club described how four trained peregrines had died after being fed pigeons shot over fields of freshly-sown corn. Another pair became ill after eating a deep-frozen pigeon, shot at the time of the spring sowing. The failure to feed and lack of co-ordination, with a tendency to press their heads on the ground or into a corner, added to the weight of circumstantial evidence that the deaths were caused by dieldrin poisoning. The Ministry's Veterinary Laboratory had failed to find any evidence of disease during post-mortem examinations on some of the birds. The Conservancy forwarded these reports to the Ministry of Agriculture in late June 1960. The Ministry agreed that they indicated a serious problem, but they were all based on very circumstantial evidence.

At a meeting of the Nature Conservancy committee in April 1960, Lord Hurcomb said it was high time that 'the nation's scientists' investigated more closely the circumstantial evidence linking these seed-dressings with exceptionally heavy bird mortality. It was a problem of 'great public importance, which was at present getting practically no attention'. Whilst the Conservancy had set up a Toxic Chemicals and Wild Life Section, it would be some time before its scientists had the evidence to prove whether the seed-dressings were responsible for the bird losses. The Chairman of the Conservancy was instructed to write to the Minister for Science, conveying its 'deep concern' over the effects of toxic chemicals on wildlife. The letter warned that, at the first opportunity, the Conservancy would press for statutory controls over the use of pesticides in order to improve the way in which pesticide use was kept under surveillance. There should be a complete ban on those pesticides proven to be harmful to wildlife. Lord Hailsham copied the Conservancy's letter to other relevant ministers, but affirmed that he could take no further action until the report of the (Sanders) Research Study Group had been received and studied.[6]

The Joint Committee

In an urgent attempt to gather the kind of data which would convince even the most sceptical that pesticides were having a damaging effect on bird life, the British Trust for Ornithology (BTO) and the Royal Society for the Protection of Birds (RSPB) set up a Joint Committee in August 1960. Most of the initiative came from two of the leading figures in these voluntary bodies, Stanley Cramp and Peter Conder. Norman Moore and John Ash, of the Game Research Association, attended the meetings of the Joint Committee as observers and took a very active part in its deliberations. During 1960 and 1961, the Committee organized a network of observers throughout Britain, thereby overcoming the earlier shortage of personnel to record and send in corpses found in the field. The RSPB bore the cost of engaging a London firm of public analyst and consulting chemists to carry out post-mortem analyses. It cost £5 per bird—about as much as having the organs of a human being analysed.

For all their obvious limitations, the studies carried out by the voluntary bodies succeeded in highlighting further the scale and extent of mortality during the sowing seasons 1960–1961. In its first report, covering the first half of 1960, the Joint Committee gave details of 67 incidents involving bird mortality. According to one commentator, the report was the 'first absolutely hard evidence that certain chemical seed dressings' were responsible for the widespread deaths; the chemical analyses had provided 'the missing link' between the dressings and bird deaths, to which 'Authority in its several guises' had drawn so much attention (Campbell 1961). In the case of six specimens examined, the analyst had 'no hesitation whatever, in saying that . . . dressed wheat was the cause of their death' (Cramp and Conder 1961).

By the time this report was available in February 1961, further deaths were being reported. Between 18 February and 31 July 1961, the Joint Committee received reports of 324 deaths, where circumstantial evidence pointed to the cause being toxic chemicals. Of the total, 292 were attributed to seed dressings. Incidents were recorded in 45 counties, chiefly the Eastern and Home Counties (Fig. 5.2). About 50 species of bird were affected, together with 12 types of mammal. Wood-pigeons were the most common species to be found dead, followed by pheasants and house sparrows. Although seed-eating species were the most

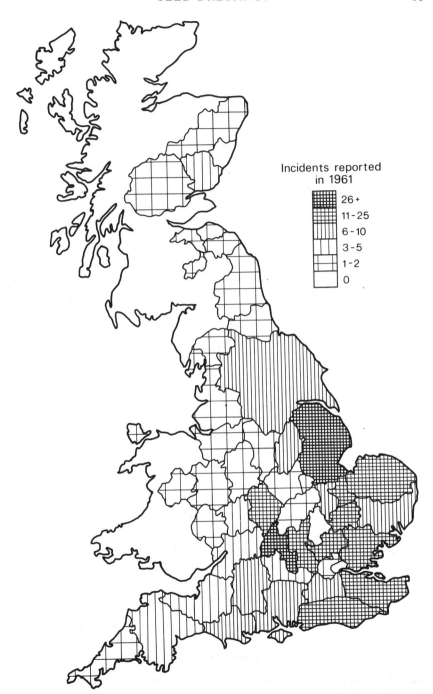

Fig. 5.2 Number of incidents reported due to toxic chemicals, 1961.

seriously affected, the corpses of other groups were found, probably
poisoned as a result of eating insects or carrion in which residues had
accumulated (Cramp, Conder, and Ash 1962).

Gradually the pattern of mortality began to emerge with greater clarity.
Fig. 5.3 summarizes the relationship between bird mortality on 29 estates
and sowing programmes on 12 of them. Because the weather was so wet,
very little autumn and winter seed was sown in 1960, and only five incidents were reported of large numbers of birds being found dead. When
sowing resumed in mid-February, most farmers not only had large
quantities of seed still to be planted, but found themselves having to drill
the corn into seed-beds that were both rough and lumpy. This led to large
quantities of dressed seed being left exposed on the surface and, even
when buried, the delay in germination caused by the dry spell from

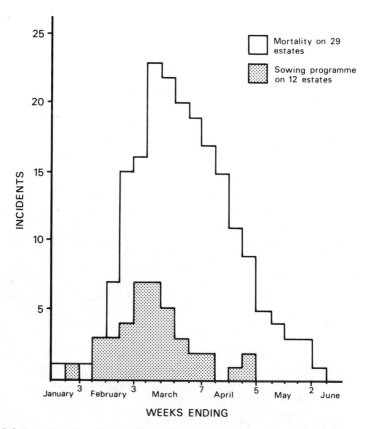

Fig. 5.3 Relationship of mortality on 29 estates to sowing programmes on 12 of
the estates in 1961.

February until mid-May caused further seed to remain comparatively accessible to seed-eating birds for long periods of time.

The number and scale of the incidents increased dramatically in 1961, but the Joint Committee had no way of estimating from the data available the total number of bird fatalities that had actually occurred over the whole country. There was every reason to believe that the destruction of birdlife was significantly under-recorded. Despite wide press coverage, many people had not bothered to report dead or dying birds found in the countryside. The number of incidents recorded in each area reflected to some extent the number of people interested in the problem, and the Joint Committee believed many totals would have been much higher if naturalists had made a more thorough search of even their own parishes. Moreover, many observers tended to look only on more open ground and for larger birds. Not only did most birds die in or under cover, but smaller corpses usually disappeared without trace after a day or two. Thus, most observers missed seeing a large proportion of those that had died. These deficiencies could only be overcome by engaging a full-time staff of qualified investigators to cover the whole country at frequent intervals. Such an approach was well beyond the resources of any organization at that time.

One of the largest kills notified to the Joint Committee took place at Tumby in Lincolnshire from late February to early June 1961. The estate comprised some 4000 acres, of which 2500 acres were arable and the rest woodland. On the basis of a detailed survey carried out by Dr J.S. Ash of the Game Research Association, it was estimated that the following had died in the woodland: 5668 wood-pigeons, 118 stock doves, 59 rooks, 89 pheasants, 16 partridges, 1 tree sparrow, 6 chaffinches, 1 starling, 5 skylarks, 1 blue tit, 21 greenfinches, 4 blackbirds, 4 moorhen, 5 sparrowhawks, 14 long-eared owls, 5 tawny owls, 4 magpies, 3 carrion crows, 2 racing pigeons, 12 unidentified birds, and a badger. The small birds, pigeons, and predators had died first, and later increasing numbers of game birds died. There was also evidence of the pheasants' eggs failing to hatch.

During the spring sowing season, 69 corpses from 32 incidents were submitted by the Joint Committee for pathological examination. In the case of 56 of these birds, there were symptoms of their having died of poisoning. A further 65 corpses were examined by a firm of public analysts and consulting chemists. In all but one bird, there were residues of organochlorines or mercury, or both, in the flesh and in such organs as the bird's crop, liver, and intestines. The levels were high enough to suggest they had been the cause of death.

In its report for 1961, the Joint Committee stressed that these heavy casualties were being sustained just before or during the breeding season, when they were most likely to have the maximum effect on the population. It was, therefore, hardly surprising that evidence was accumulating of a serious decline in the numbers of certain species, most strikingly of the peregrine falcon, kestrel, sparrowhawk, and stock dove, in the eastern and southern counties, where the greater use of toxic seed-dressings made the possibility of direct and indirect poisoning more likely. The exact significance of the many reports of reduced numbers could not, of course, be substantiated without a great deal of careful and long-term work, such as that being carried out by the BTO through the Common Bird Census. With deaths occurring on the scale reported, however, it would have been folly to delay taking any action until irrefutable proof of the decline in bird numbers and the effects of the chemicals became available. The Joint Committee warned that some species might, in the meantime, decline beyond hope of recovery.

Fox-death

It was not only the naturalist who was appalled by what was happening in the countryside. The unexplained deaths of unprecedented numbers of game birds and foxes (*Vulpes vulpes*) aroused serious concern among followers of the field sports of shooting and fox-hunting. A working alliance began to evolve between the nature conservationist and sporting interests. Monks Wood was, however, still in the planning stage. Although this alliance gave added weight to the Conservancy's proposals, a great deal of time would elapse before the Toxic Chemicals and Wild Life Section would be in a position to help explain these mysterious deaths. During that time, the Conservancy's response had to be limited and piecemeal.

Although there had been reports since March 1959, the first substantial records of what came to be known as 'fox-death' were reported in November of that year by the Fitzwilliam Hunt, near Oundle in Northamptonshire. Foxes were not the only animals to die—pigeons, cats, and a selection of other creatures died in a similar fashion. Corpses were sent to various laboratories for analysis but, as Lord Fitzwilliam admitted later, 'all our investigations led us nowhere'. The pigeons probably died from eating dressed corn and from 'some mysterious ailment or substance causing haemorrage of the lungs'. The foxes appeared to die from a disease or virus, possibly a liver disease brought on by eating the pigeons.[7]

Another Hunt reported finding 'dopey foxes' in late November 1959, and one of the corpses was found to contain traces of organochlorine. Everything seemed to point to it having fed on pigeons poisoned as a result of eating dressed seed. However, the manufacturers of dieldrin, the suspected dressing, denied such a possibility, and an examination of corpses from further areas where fox-death was reported appeared to suggest that the losses might be caused by acute hepatitis, a form of jaundice, highly infectious and known to be common in America.

The situation was regarded as being extremely serious. The Fitzwilliam country was largely denuded of foxes, and the season was just beginning when dog foxes travelled long distances. Not only might the epidemic spread further in this way, but there was a possibility of some people deliberately introducing dopey foxes to unaffected parts in order to rid the country of foxes. The Land Agent for the Sandringham Estate in Norfolk was one of several to write to the Nature Conservancy, seeking assurances that someone was investigating the outbreak. His main concern was that the mortality among pigeons and foxes might spread to gamebirds.

Perhaps the most serious aspect was the fact that no one had the responsibility for investigating the causes and extent of the outbreaks. An obvious body was the Game Research Station, particularly in view of its earlier studies of the effects of toxic sprays on pheasants and pigeons (see p. 23), but it had recently been subjected to severe economic cuts and had been forced to abandon its modest pathology unit. The Ministry's Infestation Control Laboratory agreed that the Hunts had cause for alarm and that the outbreak of disease among a large mammal population called for study, but argued that the problem was veterinary in character and best investigated through the normal private veterinary organizations. Any survey was best done by the British Field Sports Society, and perhaps the Mammal Society of the British Isles.

As for the Nature Conservancy, the Director-General admitted that 'we are helpless to do much as our scientific unit to cover these matters is only just being planned and we are unlikely to be able to bring it into full operation for a couple more years'. Even then, the veterinary and pathological work would have to be done elsewhere. The Conservancy believed there was fairly definite evidence of the fox and pigeon deaths being coincidental but not connected in any way. It was likely that the fox epidemic would continue until early summer and then die out for at least a time. There were records of large numbers of foxes having died of the 'mange' in 1895–97. It was, however, important to follow the course of the epidemic in order to provide a basis for comparison in any future outbreak. To this end, the Conservancy's regional officers were asked to submit relevant details.

In late January 1960, the Masters of Fox Hounds Association set up a committee to investigate the problem and questionnaires were sent to all Hunts. The Animal Health Trust agreed to carry out research from a veterinary angle, especially concerning the possible transmission of disease to domestic animals. In March, the Conservancy convened a meeting of interested parties to review the evidence obtained and to decide future action. The meeting did little more than emphasize the complex nature of the problem. The representative of the Animal Health Trust conceded that the cause of the disease had not been proven beyond doubt, but post-mortem examination of up to 80 corpses indicated an affinity with dog virus encephalitis, with lesions resembling those found after an epidemic on an American silver fox farm in 1930. Dr D.K. Blackmore of the Royal Veterinary College agreed that the cause was a virus, but it did not behave like dog hepatitis or fox encephalitis when in tissue culture. Spokesmen for the MFH Association were puzzled because, whilst the disease spread very rapidly, it did not always behave like an epidemic. Several isolated cases had been reported from an 'island' hunt, the West Norfolk, and the disease usually affected dog foxes only.

By the end of March 1960, news of the cancellation of hunts and often-exaggerated stories of plagues sweeping through the fox population had aroused considerable press comment. There was intense speculation as to the cause of death. A member of parliament asked the Minister of Agriculture whether the similarity of the fox disease to American fox encephalitis had caused him to review the quarantine arrangements for animals kept on American air bases in the East Midlands. The Minister replied that they were as strict as at any other airfield or port. Writing in the *Observer,* Richard Fitter sought to exonerate 'our American friends' from having introduced the disease, but he had to admit that 'we still do not know what the disease is'. It seemed to be a new virus or mutation of an existing virus. So little was known about virus diseases that every possibility had to be considered. Had a mutation been caused by radio-active fall-out from nuclear explosions over the previous few years?[8]

Any rigorous investigation was likely to fall into two parts. The Royal Veterinary College had a team of three carrying out pathological studies, and the Conservancy decided to appoint J.C. Taylor, who had just become warden-naturalist in the Breckland, to collect and collate all relevant field evidence. As Taylor commented in his report of August 1960, a lot of relevant information had already passed unrecorded. First-hand witnesses found it difficult to recall details of what they had seen, and it was hard to differentiate from their accounts between fox-death and the more usual causes of death among foxes.

The very act of announcing that someone was collecting data caused some witnesses to come forward. A Conservator in the Forestry Commission recounted how one of his staff in the Doncaster area had come upon a fox 'standing shivering violently and unable to escape and do more than snarl feebly at passers by'. A pattern soon emerged from such reports. By April 1960, large parts of the East Midlands and East Anglia were affected; the MFH Association estimated that 1300 foxes had perished, representing a substantial reduction in population.

It also became clear why so few sick foxes had been available for study in the laboratory. The sickness was so brief that most died within a few hours—the corpses were rarely emanciated. The animals became hyper-sensitive to noise and suffered convulsions, interspersed with phases when they appeared to be in a coma and partially blind, wandering aim-lessly about in daylight and devoid of any fear of humans. The most notorious instance was of a fox being discovered in the yard of a Master of Fox Hounds at Heythrop. Many corpses were found lying in or near water, or with turf in their mouths.

In addition to visiting various Hunts and laboratories, Mr Taylor spent some time in the Oundle area, where fox-death was first witnessed. Up to four-fifths of the local foxes had died, and reports of mortality among birds had not been exaggerated. As Taylor commented, 'you can't really blame people for thinking' that there was a connection between the deaths of the foxes and birds. In mid-March, the *Shooting Times* published two letters from correspondents in Essex, who believed there was indeed a causal connection. Not only had enormous numbers of pigeons died, and traces of dieldrin been found in those specimens examined, but a local vet had asserted that a great many cats had been poisoned as a result of eating the pigeons. The post-mortem symptoms of dieldrin poisoning were similar to the first indications of an animal dying from encephalitis.

By the end of March 1960, Taylor realized that fox-death had occurred almost exclusively in arable areas where large numbers of pigeons and other birds had died. Its incidence corresponded with sow-ing time, with two distinct waves in some parts coinciding with winter and spring drilling. The suddenness and nature of death resembled that recorded a few years previously during studies on the effects of dieldrin dressings on birdlife. The deaths were not confined to foxes: stoats, weasels, and cats also succumbed. In Taylor's view, the fact that the Royal Veterinary College and others could not isolate the virus or provide evi-dence of the disease being transmitted from one animal to another, seemed to confirm evidence that the foxes had died from scavenging pigeons contaminated by such chemicals as dieldrin.

The Conservancy agreed that whilst these findings were not conclu-sive, the evidence strongly suggested that the deaths were caused by the seed-dressings, rather than by a virus. When Taylor's report was circu-lated for comment in August 1960, others were less convinced. It was suggested for example that, whilst dressings might be a vital factor, the possibility of a virus disease being present should not be entirely discounted. As with other species, foxes became less immune to disease in times of stress, caused perhaps by weather conditions or lack of food. The ingestion of poisons from contaminated prey might have the kind of debilitating effect that triggered off an epidemic. The Infestation Control Laboratory had been forewarned of Taylor's conclusions as early as April 1960, and had asked the Laboratory of the Government Chemist for facilities to carry out further post-mortem analyses. The Laboratory counselled caution in attributing fox-death entirely to dressings until the results of these and other analyses were available.

The first published report to link seed-dressings with fox-death appeared in *The Times* in October 1960. An article described an experiment organized by some Essex landowner/game preservers, helped by seed suppliers. Fifteen isolated groups of 5 domestic hens were fed different brands of dressed corn, under the scientific supervision of the Animal Health Trust. A further 20 birds were fed undressed corn, and used as a control. Whereas these birds remained healthy and acted normally, at least one bird in each of the other groups died within the 30-day duration of the experiment. In 9 groups, at least 4 of the 5 birds died. The behaviour of the birds before death was equally significant. Whereas the 20 control birds laid normally, the remainder laid a total of only six eggs between them. Most of those fed on dressed corn exhibited signs of head-shaking and fatigue, similar to the abnormal behaviour patterns reported for birds suspected of having been poisoned in the wild. When some of the dead birds were fed to foxes kept in kennels, those cubs which ate them displayed symptoms similar to those associated with fox-death.[9]

The England Committee of the Nature Conservancy considered Taylor's report and these experimental findings at its October meeting in 1960. Colonel Sir Ralph Clarke, a member of this Committee and of the MFH Association and British Field Sports Society, drew particular attention to evidence suggesting that the dressings acted as a cumulative poison. Although the survivors of the 30-day experiment had been put on a normal diet, some died up to 90 days after the experiment had ended. This delayed effect increased the danger of a chain reaction between different forms of wildlife. There was a considerable risk of an apparently healthy gamebird being shot and eaten by a human being.[10]

The committee of the Nature Conservancy readily agreed to the proposal that a letter should be sent to the Ministry of Agriculture, warning of the dangers to human and wildlife and asking for the withdrawal of the dressings as a matter of urgency. With its reply, the Ministry forwarded a copy of a memorandum that had previously been submitted to the (Sanders) Research Study Group. The memorandum conceded that there had been an increase in the number and extent of wildlife incidents during the 1959–60 season. There was evidence, for the first time, of predatory animals being involved. The memorandum attributed this to the greater exposure of dressed seed during dry seasons, the increasing substitution of wet for dry dressings, leading to greater quantities of insecticide being exposed on the seed, the increasing use of seed-dressings for controlling wheat bulb fly, and the wider use of heptachlor as a seed-dressing. In an attempt to secure more information on what really happened, the Infestation Control Laboratory had embarked on a pigeon feeding trial.[11]

•An appendix to the memorandum described how 12 pigeons had been fed known quantities of dieldrin in laboratory conditions and, on death, the liver, kidneys, and flesh of the breasts, had been removed from the corpses and examined by the Laboratory of the Government Chemist. It was found that one day's consumption of seed, dressed with the compound at the standard rate, was enough to kill birds. It could take some days before a bird died after taking a fatal dose. As others had found before, the dissection and visual examination of the corpses revealed no characteristic signs of death due to the ingestion of dieldrin. Where feeding trials achieved a break-through was in indicating, for the first time, that chemical analysis could, in future, prove that a bird had consumed poison *and* state, with a fairly high degree of certainty, whether it had died as a result of consuming that poison.

Further studies were in progress to ascertain more precisely how much dieldrin was required to kill a bird, whether birds would die more rapidly if they ingested 'excessive doses', and whether small doses taken over a number of days, as might occur in the wild, had a cumulative effect. There were plans to repeat the feeding trials, using aldrin and heptachlor.

In its letter to the Conservancy, the Ministry argued that it would be premature to withdraw the seed-dressings from use in view of the programme of research that was underway and the admitted value of these dressings as a weapon against serious pests. Instead, the maximum co-operation of manufacturers should be sought to ensure the safe use of the dressings. To that end, there had already been preliminary discussions on ways of publicizing risks to wildlife.[12]

There the matter rested. Taylor's report formed the basis of a long section in the Conservancy's Annual Report. Shortly before its publication in December 1960, the Conservancy learned that the Infestation Control Laboratory now accepted that fox-death was the direct result of using seed-dressings.[13] In March 1961, Taylor and Blackmore (1961) published a short note in the *Veterinary Record*, warning that fox-death would occur again whenever birds had access to dressed corn. This was most likely to occur where alternative food was in short supply and the seed was to be found on or near the surface as a result, perhaps, of sowing conditions. Although fewer dead foxes might be found, this would be due to the reduced population, and the overall effect was likely to be even more serious.

The circumstantial evidence linking fox-death with dressed seed was eventually confirmed by post-mortem analyses of corpses found in the field and by the experimental feeding of 30 wild foxes (Blackmore 1963). In May 1960, the Laboratory of the Government Chemist had succeeded in devising methods for detecting dieldrin, and post-mortem analyses soon revealed large amounts in some carcases. The behaviour-pattern and death of foxes fed with dieldrin and heptachlor were similar to those of foxes known to have died of fox-death, and the lesions produced experimentally by administering the pesticides resembled those found in wild fox carcases at the time when dressed corn was sown. The paper concluded that the high mortality rate of 1960–61 had been 'undoubtedly linked with the use of chlorinated hydrocarbon preparations in the form of commercial seed dressings'. Insecticide poisoning could be suspected if more than 1 ppm of dieldrin or 4 ppm of heptachlor epoxide were found in the flesh and it was confirmed that dieldrin could accumulate in the body over a long period.

The first Shackleton debate

In its Annual Report of 1961, the Conservancy described the effects of pesticides on wildlife as 'one of the greatest problems facing conservationists today, not only in Britain but throughout the world'. The problem was rapidly taking on a world-wide dimension. At the Eighth Technical Meeting of the IUCN, held in Poland in June 1960, 12 papers were given on the theme of 'The ecological effects of biological and chemical control of undesirable plants and animals'. Coming from all parts of the world, most speakers expressed concern at the increasing number of sprays being introduced, their diverse effects, and the biologists' inability to forecast their side-effects. Numerous instances were cited of damage to wildlife, including such delayed effects as a reduction in reproductive capacity in birds. The main hope lay in the adoption of a combined biological and chemical approach to control, which might avert the need for more powerful chemicals and help to minimize the chances of resistant strains or new pests appearing. In the shorter term, great stress was laid on improving machinery used in spraying, and on educating the public not to use poisons indiscriminately (Kuenen 1961).

The growing interest on the part of press and public in the safety aspects of new pesticides was illustrated at a press conference convened by the Ministry of Agriculture in February 1961. Although the conference was intended to launch a new Agricultural Chemicals Approval Scheme, an observer from the Nature Conservancy reported that, in their questions, press correspondents showed just as much interest in the safety aspects of using pesticides and, therefore, in the Notification Scheme. The Conservancy believed this augered well for the time when it would be ready to make its own detailed proposals for regulating the use of agricultural chemicals.[14]

There were references to the pesticides question in parliament. In a debate on scientific policy in November 1960, Lord Shackleton drew attention to the development of new pesticides, the long term effects of which were unknown. His remarks were not taken up by the Minister for Science, who replied to the debate on behalf of the Government. In the following March, Viscount Elibank put down a Question drawing attention to the report of the Joint Committee of the BTO and RSPB and asking what further steps the Government was taking to collect evidence. Other members drew attention to public misgivings as to the apparent lack of concern on the part of the Government. On 25 April 1961, Lord Shackleton initiated a major debate by putting a further Question asking whether the Government was 'aware of the catastrophic casualties caused to game and wild life, especially during the last few weeks, which are attributable to the use of toxic chemicals in agriculture, and to ask what immediate steps they propose to take to deal with this urgent problem'.[15]

In a maiden speech, Lord Walston appealed to members not to be too dismissive of the problems arising from the death of some pheasants and foxes. Although such losses might be petty in themselves, they were symptomatic of a trend so serious that it called for positive intervention on the part of the Government. Viscount Goschen stressed that the pesticides issue was unusual inasmuch as 'no one is fighting anyone else'. He continued:

the farmers want to use aids for their production, and they want to use aids which are economical, but they have no desire to kill off the bird life or animal life of the country. The people who produce these toxic chemicals have no desire to kill off the wild life of the country. The Government are anxious that all parties should do well . . . the general public want to see the bird life and animal life protected; and they want to see the farmers happy . . . so that really no one is against anyone else.

For his part, Lord Shackleton conceded that the subject was exceedingly complex, and that there was nothing to be gained by scaremongering or attacking farmers and manufacturers. Chemicals had played a major part in increasing agricultural output. It was not enough, however, to say that man had always interfered with nature, and that most farmers and manufacturers were responsible people. The use of chemicals had now led to a very dangerous situation; for the second year running, heavy casualties among bird and mammal life were being reported during the spring sowing season. In causing such a catastrophic change, the agricultural chemist had, Lord Shackleton said, achieved a record that would have been the envy of the Borgias.

Lord Shackleton summed up the Government's attitude as being 'this may be a serious problem, but we cannot interfere with processes of agriculture and we hope that everything will be all right'. A well-known chemical supplier, he continued, had been quoted as saying it was not his responsibility to investigate the side-effects of his chemicals. The manufacturers had done nothing in the way of long-term ecological research and the public and Government would probably still be unaware of the seriousness of the situation were it not for the wildlife societies and public-spirited individuals.

Shackleton said that if Britain could criticize the destruction of game in Africa, it was incumbent on her to make sure her own hands were clean. We should not, he said, continue to destroy irresponsibly the heritage of this and the next generation 'in the name of freedom or anything else'. Quite apart from aesthetic and humanitarian reasons for curbing the use of pesticides, there was a real economic risk arising from the extermination of wildlife. In his opinion, there was no time to wait for further committee reports. A ban should be imposed on all dual-purpose seed-dressings that contained dieldrin and no chemical should be used unless a body like the National Agricultural Advisory Service deemed it essential. If it proved exceptionally dangerous to wildlife, it should be withdrawn. Above all else, the Government should provide adequate funds for long-term research rather than the miserable amounts presently available.

In his reply to the debate, the Joint Parliamentary Secretary to the Ministry of Agriculture, Earl Waldegrave, emphasized that the Ministry was wholly against 'particularly dangerous kinds of dressing being used wholesale, whether required or not'. There was, however, a risk of emptying the baby out with the bath water, and of perfectly innocent chemicals being condemned. In view of their benefits to agriculture, it was important not 'to jump to unjustifiable conclusions'. Whilst some chemicals might have a harmful effect, there was no evidence that they could cause a catastrophe. Earl Waldegrave continued:

what we have to do is to minimize that (danger) to the greatest extent, while not denying the immense value to agriculture of these chemicals, properly used and applied.

It was, he concluded, largely a question of timing: it was wrong to expect an instant solution to so novel and complex a problem.

The Select Committee on Estimates

Both before and after the debate of April 1961, the Ministry of Agriculture was under considerable pressure from another, and wholly unexpected, quarter. It so happened that the House of Commons Select Committee on Estimates had decided to scrutinize the Ministry and, from January until mid-May, a sub-committee took evidence on all aspects of its work. In February, the Conservancy learned that the sub-committee intended to investigate the ecological effects of 'the use of toxic spraying, pesticides, seed-dressings, etc., approved by MAFF', in order to assess how far the Ministry was implementing Government policy, and carrying out its statutory duties 'as economically as possible' and with sufficient regard to the advice of 'outside scientists and other authorities'.[16] The sub-committee felt the question was so important that it devoted a fifth of its report to the use of toxic chemicals (Select Committee 1961).

The report set out the evidence for a substantial loss of wildlife. The first intimation came from a witness known personally to the Chairman and another member of the sub-committee—the kennel-huntsman of the South and West Wiltshire Hunt. He recounted stories of pigeons dropping dead out of the sky, and of foxes, badgers, and cats dying from having eaten these birds. He described the new seed-dressings as 'a serious menace to all forms of wild life'. Even man was not immune. The witness related how a farmworker had caught a pigeon and intended to have it for supper until the farmer suggested he might open the bird's crop. The labourer changed his mind when he discovered the crop full of dressed corn. The Chairman of the committee told the witness, 'you have shaken us very much'.

The game and nature conservation bodies convinced the committee that incidents were both serious and widespread. The Game Research Association described them as ranging from the death of a few birds to that of hundreds and thousands. The Nature Conservancy estimated that the number of cases reported in 1961 was three times greater than that in 1960. A witness for the Council for Nature asserted that 'there has been no parallel in the present century, or at any time so far as I am aware, to the recent drastic reductions in the number of such birds of prey as kestrels and certain species of owls, quite apart from game and certain other birds'.

From the committee's point of view, the most disturbing evidence came from those landowners, farmers, and surveyors who had been called to give evidence on other aspects of the committee's inquiry, and who could not be regarded as being unduly biased towards nature conservation. A farmer of 3000 acres in the East Riding was called to give his views of the agricultural advisory services, but he nevertheless expressed his very serious concern over the side-effects of the newer seed-dressings. During the previous two years, he said, 'a tremendous lot of dead birds had been picked up during the spring sowing season', including game and those species beneficial to farming. Although such evidence was anecdotal and based on impressions, members of the committee were shocked by the unanimity of feeling among the witnesses.

By the time the committee heard evidence from witnesses for the manufacturers, members were profoundly sceptical of their claim that the deaths of 1960–61 were lower than those of the previous four years, and that every step was being taken to safeguard wildlife. The manufacturers argued that the increase in the number of dead birds reflected the way in which fields and hedgerows were 'being combed by devoted people'. Not only had dead birds always littered the countryside, but many of those recently found had died from other causes. The manufacturers reminded the committee that seed-dressings were worth £11 000 000 per annum in agricultural production, and that no agricultural chemical was completely free from potential, and sometimes real, danger. It was 'a tremendously difficult job to find things that are sufficiently specific to do the job you want without danger'. Witnesses stressed that the industry was seeking alternative chemicals but, as they emphasized:

you will understand the difficulty of being able to direct research to discover things: one cannot suddenly discover things because they are needed. There is quite a lot of work involved.

The Chairman of the committee, Sir Godfrey Nicholson, warned the witnesses that 'Parliament and the public will not just be content to be told that investigations are continuing and that of course this is a very long and expensive job'. No one pretended that it was easy to solve the problem. He continued, 'this is a big problem which cannot be written off with generalities, however sound those generalities may be'.

In its written evidence to the committee, the Conservancy identified six ways in which the necessary facts might be obtained about the effects of chemicals on wildlife. Industry was already largely responsible for measuring the tolerance of representative animals to the chemicals (the laboratory LD_{50} tests). Likewise, the Government Chemist was involved in discovering reliable methods for diagnosing the cause of death where animals were thought to have been poisoned. The role of the Conservancy was to provide background data on the species affected, and to initiate long-term research. And it was for the Ministry to obtain, from field and laboratory work, the direct evidence on the effects of pesticides. In almost every case, the resources to discharge these various responsibilities were totally inadequate.[17]

In the course of giving evidence, the Permanent Secretary of the

Ministry, Sir John Winnifrith, agreed that it was the Ministry's responsibility to protect human and wild life from the injurious effects of pesticides used in farming. The committee attributed the fact that the Ministry had been 'a little taken by surprise' by the loss of wildlife to its failure to collect comprehensive and accurate information on the effects of toxic chemicals. The only field and analytical evidence available to the committee had come from the voluntary bodies. Some of this information was relevant not only to wildlife but also to the preservation of human life. The Ministry was in no position to confirm or deny the findings of the voluntary bodies.

It was not until October 1960 that the Ministry had instructed the Infestation Control Laboratory and its Regional Pest officers to investigate the circumstances of any incident reported to local officers. It then stipulated that, where appropriate, the corpses were to be sent to the Laboratory for post-mortem analyses. The committee regarded this support as too little and too late. The sowing season was virtually over by the time the public knew of the arrangement, and the response to the request for information had been poor.

Furthermore, not only had the Ministry failed to fulfil its potential role in gathering data, but the lack of information had caused it to overlook problems arising from the use of toxic chemicals. The committee regarded the requirements laid down by the Notification Scheme as 'gravely inadequate'. Manufacturers operated in a highly competitive world, where speed and secrecy were overriding factors when considering whether a product should be marketed. They carried out 'modest tests on a very few kinds of wild life', and assumed that all other species would react in the same way. The committee supported the Nature Conservancy in urging the Ministry to extend the Notification Scheme so as to oblige manufacturers 'to conduct comprehensive tests and field trials in a manner and according to standards laid down by the Ministry'.

In evidence to the committee, the Ministry laid great stress on the number of warnings issued with regard to the dressings. Sir John Winnifrith read an extract from a Ministry booklet, warning that:

dressings containing dieldrin, aldrin and heptachlor can kill birds that eat treated seed. Great care should be taken not to leave any treated seed lying about when it is being stored or sown. Higher strength dressings for wheat bulb fly should be used only on winter wheat and then only in areas where there is real danger of attack.

Witnesses for the manufacturers stressed that they relayed such warnings to their seed merchants and farmers.

Despite such assurances, there was plenty of evidence that these warnings lacked impact or were irrelevant. Farmers found it difficult to dress their corn in the right quantities, and often added an extra dose in order to make sure all the grain was treated. Fortunately, over 90 per cent of the seed supplied to farms was ready-dressed but, even here, there was scope for misuse. The committee heard of dressed seed being poorly labelled, without instructions on how the seed and containers were to be handled. The fact that the constituents of the dressings could be changed without the user being informed made farmers uneasy when game and wild birds on their land suddenly died without apparent reason.

The committee regarded arguments as to the efficacy and timing of warnings as largely irrelevant in view of the fact that some birds would always find the toxic seed, no matter what precautions were taken. The most elementary field-trials should have made this clear to the industry and farmers. There was the old adage that farmers planted one seed 'for the crow, and one to grow'. There was a particular risk of birds probing the soil for seed in January and February, when alternative food was scarce. They would also find more grain in wet seasons, when the bad soil conditions made drilling difficult. The Game Research Association believed the prevalence of such conditions in 1960–61 had contributed to the high bird mortality.

A sub-committee of the House of Commons Estimates Committee was not the most obvious forum for a recommendation as to whether a type of agricultural practice should be discontinued in the interests of nature conservation. The committee was, however, very alarmed lest 'the caution and leisure of the Government machine' should lead to 'another series of tragedies next spring and possibly there being very little wild life left'. Sir Godfrey Nicholson told the witnesses from the Nature Conservancy that 'the primary purpose of this Sub-committee—the only purpose really—is to suggest economies in expenditure', but it is always possible to advise that short-term expenditure should be incurred if it led ultimately to long-term economies. He therefore asked them if the damage being wrought 'to the ecology of this country' was so great that expenditure to arrest it would lead to economies in the long run. The Conservancy gave him that assurance.

It was in this context that Sir John Winnifrith was examined during his last appearance before the committee. The Chairman asserted that the Notification Scheme had broken down, in so far as it had failed to prevent birds from eating the poisons. The Permanent Secretary was pressed to say what steps his Minister would take to prevent the same thing happening during the next sowing season. Sir John refused to say, arguing that there might be other reasons for the exceptional mortality. There was at present, he said, no scientific way of proving the validity of the estimates of mortality that had been given by the Game Research Association and other witnesses. Any figures had to be set against the total bird population, which was unknown. In his words, 'what you want to know is what proportion of the population is being wiped out'. From the chair, Sir Godfrey Nicholson replied:

of course I accept that, but I think you would also accept the fact that if you are going for 100 per cent scientific accuracy a good deal of time is certain to elapse, and I am really appealing to you, not as a scientist or an administrator but as an ordinary man of commonsense, in which case I think that the evidence, morally speaking, is irrefutable, and I do hope, beg and pray the Ministry will not let bureaucratic caution let another sowing season elapse with these horrible consequences. This past weekend I have seen corpses myself, and I do not like it. I do not want to fear that I, as a Member of Parliament, or this Sub-Committee have been neglectful in urging on the Ministry what I feel should be their duty.

In its report, published in June 1961, the committee argued that 'sufficient scientific and circumstantial evidence' was available to prove that 'these exceedingly toxic chemicals' had caused the recent mortality of wildlife. As well as wood-pigeons and other agricultural pests being affected, a number of species beneficial to farming had also suffered. Although the value of dieldrin, aldrin, and heptachlor was beyond doubt, restrictions should be imposed on their use.

The voluntary ban on seed-dressings

Throughout 1961, the Conservancy had pressed the Advisory Committee on Poisonous Substances to exert far greater control over the actual application of chemical sprays. In the same way as a farmer had to follow a prescribed procedure before receiving a ploughing grant, so should he be required to forward details of his proposals for spraying, including such information as treatment dates, operatives, and chemicals to be used. The approval of Ministry officials should be obtained before spraying took place and only skilled operators should be allowed to apply the chemicals. They should be trained in the value, dangers, and uses of toxic chemicals, and be licensed by the Ministry or the County Agricultural Executive Committees.[18]

These points were reiterated in a letter of June 1961, in which the Nature Conservancy committee pressed the Ministry of Agriculture for a complete ban on the use of aldrin, dieldrin, and heptachlor. At the very least, the use of these chemicals should only be permitted when authorized by the Ministry. Although there was still much to learn about the effects of individual chemicals on wildlife, enough was known to raise serious doubts as to the safety of these particular pesticides. The products should be placed in a 'danger category', and treated very much like poisons in the pharmaceutical world.[19]

Once again, the Chairman of the Conservancy wrote to the Minister for Science, describing how the situation had deteriorated further. It was now possible to speak of 'a mass killing of birds' at the most critical phase of the year, when breeding was underway or about to start. There was also disturbing evidence of the poisons being transferred in lethal doses to other species, such as fox and peregrine falcon. Although full and precise scientific assessments on many of the finer points were still lacking, it was clear that wildlife was being affected on a large scale, inflicting considerable cruelty on the animals involved. The letter reminded the Minister of how the Conservancy's warnings in 1960 had proved fully justified, and the Chairman asked for support in the forthcoming discussions with the Ministry of Agriculture. The Chairman wrote:

while of course we realise the advantages of acting so far as possible in agreement with the Manufacturers, the case for Government control of toxic chemicals is, in the Conservancy's view, overwhelming and we suggest that a minimum immediate aim should be the prohibition of the use of specified seed dressings next year.[20]

The drafting of this letter was the occasion of a debate within the Conservancy over whether it was prudent to press for a ban on *all* seed-dressings. There were fears that such a course might lead to the Government deciding that the loss of wildlife was preferable to the real (or imagined) loss of crops. Instead, it was decided that a different approach should be adopted, whereby the Conservancy should press for the use of seed-dressings to be prohibited, except where severe crop damage would be the inevitable consequence of their not being used. Dr Moore cited such cases as that where old pasture was ploughed up and sown for the first time, or where land was known to be infested with wheat bulb fly. He forecast that the repercussions for wildlife of this near-total ban might, in practice, be little different from that of total prohibition, and would help to make the controls more palatable to the farming industry—and therefore more effective.[21]

In its report, completed on 28 June 1961, the Estimates Committee supported the Conservancy and other natural history bodies in recommending that the Minister and the Secretary of State for Scotland should secure the immediate prohibition of seed-dressings containing aldrin, dieldrin, and heptachlor, and other chemicals of comparable toxicity to wildlife. On the next day, the Ministry convened a meeting of interested parties to discuss the effects of pesticides on wildlife. Once again, it was agreed that there was no certain proof that the chemicals actually caused the deaths, but the circumstances in which the bodies were found left little room for doubt, especially as some of the birds were found to have consumed seed dressed with the chemicals shortly before they died. This was particularly striking among birds analysed in the springtime. The

Conservancy pressed for a complete ban on the use of dressings containing aldrin, dieldrin, and heptachlor, or, at the very least, for the dressings to be permitted only under specific instructions from an officer of the Ministry.

At the meeting, the Ministry made an alternative proposal, which was eventually accepted by all parties, namely to restrict the use of these seed-dressings to autumn and winter sowings where considerable damage from wheat bulb fly was expected. Recent studies had shown that aldrin, dieldrin, and heptachlor were not much more effective than gamma-BHC for controlling wireworm at any time of the year. Gamma-BHC was slightly more effective for controlling wheat bulb fly in early spring whilst autumn was the only time of year when there were definite advantages in using aldrin, dieldrin, and heptachlor. The findings were highly significant in view of the way birds depended on different foodstuffs at different times of the year.

Since the summer of 1960, the Infestation Control Laboratory of the Ministry of Agriculture had been conducting experiments on the wood pigeon. Using the levels of pesticide residue recorded in laboratory feeding trials as a guide, an examination was made of levels found in 91 corpses retrieved from the countryside. It was found not only that aldrin, dieldrin, and heptachlor were highly toxic to pigeons, but that there was no repellent available that would deter birds from eating dressed seed. The Laboratory concluded that the only way of reducing the very real hazards to birdlife was to modify the use of the insecticides (Turtle, Taylor, Wright, Thearle, Egan, Evans, and Soutar 1963).

A great deal of information on the wood pigeon came from a study of 2650 acres of arable farmland on the Cambridgeshire/Suffolk border. Close observation indicated that the pigeons only fed on autumn sowings when the better grain supplies of the stubble were not available. It was unusual for more than half the population to feed on them. The spring-sown corn might sustain over 80 per cent of the population (Murton and Vizoso 1963). In view of the seasonal differences in the effectiveness of different pesticides, and the greater vulnerability of birds in spring, the Ministry concluded that there was 'not sufficient justification to warrant the continued use of aldrin, dieldrin, and heptachlor, except against wheat bulb fly in autumn'.

The voluntary ban on the use of these compounds as spring dressings took effect in the spring of 1962. It was formally negotiated under the Pesticides Safety Precautions Scheme, and was renewed in the next and subsequent years. In a Second Special Report from the House of Commons Estimates Committee in November 1962, the Ministry emphasized that the restrictions had proved extremely effective, and that the situation was now under control (Select Committee 1962). A Ministry spokesman warned industry and farmers that no relaxation in the voluntary ban could be expected. The Joint Committee of the BTO and RSPB received only 41 reports of incidents attributable to seed-dressings in the spring of 1962 (compared with 292 in 1961). The Joint Committee attributed this to the ban and the excellent sowing conditions, both in the autumn of 1961 and in following spring. The number of reports of death from other forms of pesticide use continued to be high, and it remained to be seen whether the ban would be equally effective in reducing mortality when sowing was made difficult by the weather (Cramp, Conder, and Ash 1963).

For its part, the Nature Conservancy welcomed the partial ban, but the Annual Report of 1961 warned that it was only the first, easiest, and simplest step to take. Far too many people, including some scientists, overlooked the equally serious and possibly more widespread dangers arising from the continuous absorption of sub-lethal doses of poisons by wildlife. By reducing the rates of reproduction, and by increasing susceptibility to disease, the effects of sub-lethal poisoning might be far more serious ecologically than the 'catastrophes' experienced in the springs of 1960 and 1961.

Whilst it had taken initiatives over the seed-dressings issue, and the first appointments had been made to the Toxic Chemicals and Wild Life Section, the Conservancy could provide little in the way of research facilities or data. A research capability cannot be built up overnight. That capability was, however, beginning to develop much more quickly than had been expected. The seed-dressings issue not only reinforced the need for research on toxic chemicals within the Nature Conservancy, but it provided both expertize and experience in identifying how research findings could be applied and promoted.

6

THE PERSISTENT PESTICIDES

With the dramatic decline in bird deaths, following the ban on the more toxic seed-dressings in springtime, some observers thought the pesticides problem had come to an end, and many more hoped that would be the case. The Conservancy was appalled by such an attitude. If such complacency affected the mind of Government, any hope of further extensions in surveillance and regulation would disappear and, with them, the chances of promoting further research—that is, until the next inevitable crisis. The last thing the Conservancy wanted was this kind of disruption, just as its toxic chemicals research was getting under way.

Fears that this might happen were aired by Dr Worthington in May 1961, following a meeting at the Office of the Minister for Science. At the Minister's request, representatives of the Research Councils and of the Ministry of Agriculture had been assembled to consider what action should be taken in the light of the Shackleton debate and drafts of the report being prepared by the (Sanders) Research Study Group. A spokesman for the Study Group confirmed that it was most concerned about the impact of pesticides on wildlife, but any boost which his remarks may have given to research was counterbalanced by his further observation that these anxieties were now allayed by the regulations to be imposed on seed-dressings. Dr Worthington found himself in a minority of one at the meeting in protesting that, although seed-dressings had created the greatest havoc, they were only one in a series of episodes causing damage to wildlife and that, given the rate at which new pesticides were appearing on the market, further emergencies were to be expected.[1]

The Conservancy contended that there were much more serious, long-term side-effects still to be tackled, arising from the relative stability of some pesticides in the environment. In 1960, Charles Elton drew the attention of Norman Moore to a paper, published in a little known journal, in which two members of the California Department of Fish and Game described what happened to the wildlife of Clear Lake, a body of water of 41 600 acres in California, following an attempt to control the gnat *Chaoborus astictopus* with DDD (TDE) in 1958. Hunt and Bischoff (1960) found that the residues in all the flesh samples taken in the course of their surveys exceeded the specified rate of dilution of active insecticide in the lake water. Carnivorous fish seemed to accumulate greater amounts of residue than plankton-eating fish of the same size. Fish appeared to be more tolerant than species higher up the food chain.

Losses of western grebe (*Aechmophorus occidentalis*) were particularly severe. The authors of the study attributed the decline in population from over a 1000 pairs to only 25 to the accumulation of TDE-residues in the birds.

Although the paper aroused little comment at the time, it provided Dr Moore with further evidence that the most serious side-effects of pesticide use were still to be recognized and resolved. These arose from the relative persistence of those compounds which were less readily broken down into simpler and biologically inactive substances. Whereas 2,4-D might persist in the environment for up to four weeks, gamma-BHC could remain for an estimated 14 years. Persistence was the most outstanding attribute of the organochlorine compounds. Whereas farmers preferred the less persistent pesticides for most purposes, in order to minimize the risk of high levels of residue still being present when crops were eaten, it was a great advantage to use pesticides that would remain active for at least several months when protecting seeds, young plants, or sheep from parasites. This was one of the main reasons why, by the late 1950s, increasing quantities of the organochlorine compounds were being used as seed-dressings and in sheep-dips.

It had been hard enough to persuade the sceptical and apathetic that the sudden losses of thousands of birds in 1960–61 were caused by pesticides; it was even more difficult to convince them of the more subtle dangers posed by the persistence of certain types of pesticide.[2] Almost by definition, the effects of persistence could not be appraised in laboratory tests or field trials carried out in the period before the pesticide was marketed. The effects could only be recognized as a result of close surveillance during the first years of commercial use.

The use of persistent pesticides called for a much broader approach to pesticide regulation. Whereas anyone considering acute toxicity might think that DDT was much less dangerous to wildlife than, say, the organophosphorus insecticide, demeton-methyl, which was five times more toxic, things were very different from an ecological point of view. Even if all the wildlife in the treated area was killed by demeton-methyl, the pesticide soon broke down into harmless constituents, and the area was soon recolonized by wildlife. There was not time for residues to spread more widely. Nature was used to dealing with this type of catastrophe. On the other hand, because of their solubility and persistence in fatty tissues, DDT and other organochlorines would continue to harm and kill wildlife over a much longer period. Persistence made it easier for them to be 'carried' into hitherto unaffected organisms and environments, and it was not long before residues were found in various mammals, birds, and fish in some of the more remote parts of the world (Moore 1965*a*, *b*).

Ecologists in Britain were among the first to grasp the environmental significance of what was happening. The residues were both widespread and their toxicity cumulative. They were being released at a rate faster than the environment could destroy them. The side-effects of pesticide use could no longer be regarded as a local problem. In America, there had been a number of warnings of how this might happen. A great deal of relevant research had been carried out by Robert L. Rudd for the California Department of Fish and Game and, after 1958, for the US Conservation Foundation (Rudd and Genelly 1956; Rudd 1964). During a visit in the spring of 1963, Dr Moore learned that American scientists were finding residues of DDT in samples of cod liver oil and in air particles (collected by a windsock drawn behind an aircraft). The larger implications of their individual findings had often not been appreciated.

As the work of the Toxic Chemicals and Wild Life Section moved from preliminary reconnaissance, consisting largely of field observations, towards a long term programme of field and laboratory experiments, it was decided to concentrate on studying the effects of repeated applications of persistent chemicals on different forms of wildlife. Particular attention was directed to the effects of organochlorines on animals at the ends of the food chains, with special emphasis on avian predators and their eggs.[3]

The Wildlife Panel was the obvious forum in Britain for examining the implications of this new phase in pesticide development. Appointed in 1959 (see p. 35), the Panel had drawn up an Appendix to cover the wildlife aspects of the Notification Scheme. That done, a new Panel had been formed to keep the use of toxic chemicals under general surveillance. In early 1963, a new guide was drafted for the pesticide industry to use in assessing the risks arising from the new compounds. Prepared in collaboration with personnel from the Ministry of Agriculture, it was largely based on the findings of the Toxic Chemicals and Wild Life Section. Although there was still considerable inertia to be overcome before adequate safeguards were introduced, the Panel had taken a major step forward.[4]

From this and other initiatives, it is clear that there was not only a growing body of opinion in Britain convinced that toxic chemicals brought both benefits and dangers, but that relevant research was under way by the time Rachel Carson caught the public imagination with her book, *Silent spring.*

'Silent spring'

Rachel Carson had been on the scientific, and then the editorial, staff of the U.S. Fish and Wildlife Service until 1952, when she took early retirement in order to spend all her time writing books. There followed three volumes on aspects of marine ecology and conservation, and then, in 1962, *Silent spring,* a book of 297 pages, with an additional 54 pages listing her principal sources, and an 11-page index. In this book, Miss Carson made a blistering attack on the indiscriminate use of chemicals to control agricultural pests and diseases. She claimed that the large-scale use of toxic chemicals would lead to long-term genetic and ecological problems as a result of the build up of residues in soil and food chains. There was an urgent need, she wrote, for a more thorough understanding of the effects of the chemicals, and for more stringent controls on their use.

Miss Carson was already known to the Conservancy. In November 1960, she had written to the naturalist and author, James Fisher, asking for details of the use of herbicides for controlling roadside vegetation in Britain. For several years, she wrote, 'I have been working on a book on pesticides, stressing the violence to ecological relationships that is done by their wholesale use. In my opinion the chemists and the engineers are leading us into very grave difficulties and the biologists are not making their views known with anything like the necessary effectiveness'. The letter was passed to the Conservancy, and the Director-General arranged for relevant information to be sent to Miss Carson. Mr Nicholson offered to meet her on a forthcoming visit to the United States, and he added in his letter, 'I only wish we had more people on this side who could write as you do on conservation'.[5]

Prior to the book's appearance in America, most of the text was published in three long instalments in the magazine *New Yorker* during June and July 1962 (Carson 1962). The Minister of Agriculture made his first response to the instalments when answering a Parliamentary Question. He emphasized that they contained nothing new. The greatest precautions were already taken in Britain, and no dressing or other form of pesticide was allowed without being tested and approved. Some members of the House of Commons, including Sir Godfrey Nicholson, thought the reply was far too complacent. On reading the first instalment in *New Yorker,* a member of the Office of the Minister for Science asked the Director-General of the Conservancy for an assessment. If the picture conveyed by Miss Carson was correct scientifically, the long-term outlook was very dangerous, even allowing for the much greater use of chemicals in America.[6]

Mr Nicholson was not only familiar with the British situation, but his visits to the United States since 1959 had made him conversant with developments there. He had met several of the specialists cited by Miss Carson, and he was able to assure the Office of the Minister for

Science that the information on wildlife was authentic. The status of the larger birds of prey had declined further since the book was written. Although no one could prove that pesticide residues were responsible, all the proposed alternatives seemed to have been eliminated. Officials in the U.S. Fish and Wildlife Service had found astonishing concentrations of DDT in bald eagles; they had to go as far afield as Alaska to find birds that were free from residues. Although specialists in Britain had little to learn from *Silent spring,* the book was of great value in providing a much clearer picture of the general trend. It was a warning of how the use of pesticides was likely to be criticized in the future. The conflict would probably be more acrimonious than the current one over the cancer-producing properties of cigarettes. If the Government wanted to be in a position to assess accurately the emotional and biased generalizations made by both sides, there had to be much more research.

In a minute of September 1962, Dr Moore drew attention to three important points made by Miss Carson. First, she related how the dangers posed by these totally new substances were almost exactly parallel with those arising from increased radio-activity in the environment. Mankind was 'running a new kind of risk by using these substances and an extremely critical watch should be kept on their development'. Secondly, the most serious problems were the long term genetic and ecological ones, and ecologists had a special responsibility to emphasize and investigate their significance. Thirdly, the existing administrative machinery in America and elsewhere was quite inadequate to deal with a situation which, like industrialization over a hundred years previously, had crept up on society unawares.[7]

Press articles soon appeared under the headlines, 'We are being slowly poisoned' and 'Will weed-killers one day kill us?' When the pesticide and farming industries protested, the Conservancy reminded them that, for many years, they had been making equally unbalanced statements in respect to pesticides. It was hardly surprising that some of the frustration and resentment of conservationists was now boiling over into print. Many scientists in agriculture and industry had been openly contemptuous of 'the very real problems of effects on wildlife'.[8]

The publication of the excerpts and *Silent spring* in the United States marked an important turning point in the regulation of toxic chemicals. The case for the other side had been put forward and 'accomplished extremely well'. At long last, there was a foil to the deluge of manufacturers' brochures, the articles in the farming press, and even the publications of the Ministry of Agriculture. In the opinion of the Conservancy's Deputy Director, Barton Worthington, the Conservancy could once again 'afford to sit back and be dispassionate'. The time was right for reassessment, and the Conservancy should seize the opportunity presented by *Silent spring* to secure more resources, so that an adequate scientific appraisal could be made.

Mr Nicholson wrote to the Office of the Minister for Science in this vein. In his words:

> we are entering an uncharted field: we obviously need constant vigilance but this must be imbued with a much more critical approach and an appreciation of the intricate, complex and far-reaching inter-actions stemming from the use of toxic chemicals in agriculture.

Two crucial ecological questions had to be answered, namely, how did the pesticides pass through food chains, and how were they concentrated in predators. In its search for answers, the Conservancy's Toxic Chemicals and Wild Life Section was handicapped by inadequate resources, which limited in turn its contribution to the Wildlife Panel and the ultimate success of the Notification Scheme.[9]

At the meeting of the Inter-departmental Advisory Committee on Poisonous Substances, in November 1962, Mr Boote again attacked those who sought to refute Miss Carson with statements that were equally emotional. He described how, on a recent visit to the United States, Dr Worthington had taken the opportunity to check the references to wildlife in *Silent spring*. They were found to be sound. Mr Boote emphasized once again that the Conservancy found the book and articles 'a significant warning on the long-term and cumulative effects for wild life'. Whilst it was true that the pesticides question had received considerable attention in late years, there was still need to put the Notification Scheme on a better research basis, so far as wildlife was concerned.[10]

Plans were now far advanced for the publication of the British edition of *Silent spring*, with an introduction written by Lord Shackleton, who had already played a conspicuous part in raising the pesticides issue in parliament. In preparing his introduction, Lord Shackleton looked to a range of authorities, and particularly the Nature Conservancy, for advice. The Secretary of the Medical Research Council, Sir Harold Himsworth, described how a close and continuing watch was being kept on the effects of pesticides on human health. Many of the unpleasant possibilities set out in *Silent spring* had been foreseen, but the Council saw no reason for sounding alarms. In its view, the book lacked perspective by giving 'undue attention to one end of the spectrum of possible ideas'. It was particularly difficult to prove 'a negative possibility', namely that pesticides were not harming human health. The possibilities of ill-effects had not been ruled out by the Council, but they seemed increasingly unlikely as the results of trials and experiments became available.[11]

The Parliamentary Secretary to the Ministry of Agriculture told Lord

Shackleton that the Ministry found the purpose of *Silent spring* laudable. It supported the author's final conclusion that 'chemical controls should be used with moderation and alternative methods should be developed as rapidly as possible', but, Lord St Oswald continued, the Ministry regarded the book's most disturbing conclusions as mistaken and based on inadequate evidence. The book was 'that of a skilful and determined advocate: objective assessment is at a discount in her pages'. Chemicals fulfilled an essential purpose in pest control. Another recently published book had described how, before the introduction of chemical pesticides, wireworm had destroyed 120 000 tons of wheat annually, carrot fly had caused a loss of 31 000 tons of the carrot crop, and flea-beetle damage had reduced the root harvest by 1.5 million tons (Laverton 1962). Lord St Oswald emphasized that farmers used no more toxic chemicals than were absolutely necessary. Not only did the Advisory Service and the research stations stress the need for moderation in their use, but the farmer found the chemicals very expensive to purchase and use.

The attitude of the Office of the Minister for Science emerged in January 1963, when the text of a letter was circulated in confidence to the Research Councils. The Minister had been sent an American copy of *Silent spring* and, in his letter of acknowledgement, Lord Hailsham agreed that it was a thought-provoking book. It was welcome in so far as it warned of dangers and of the need to improve the regulation of pesticide use. He regretted, however, the author's failure to stress the benefits that had been conferred on farming by the use of chemicals. Britain had escaped some of the wilder practices described in the book. The Minister's main concern was lest *Silent spring* should let loose a campaign of hate against the use of chemicals by the 'muck and magic' school of agriculturalists, who always seized on events of this kind to promote their own pseudo-scientific beliefs. Although Miss Carson was right to draw attention to biological, as opposed to chemical, means of control, these were pretty well known, and Lord Hailsham doubted if British scientists had 'missed many tricks on that front'. Following the report of the (Sanders) Research Study Group, research had been reappraised, and the scare over seed-dressings had led to a tightening-up of regulations on their use.[12]

The publication of *Silent spring* in Britain was planned for February 1963, and the various interested parties put the finishing touches to their preparations for the event. The Association of British Manufacturers of Agricultural Chemicals (ABMAC) published a review and commentary, which described *Silent spring* as 'a welcome book, from the reading of which no one should be deterred by its concentration on the darker side of the subject'. It would stimulate everyone to take even greater care in the use of pesticides, although it was important to maintain a sense of perspective. The problems of world health and food supplies were even more pressing than many of the issues covered by Miss Carson (ABMAC 1963).

Many reviews of the book were exceedingly critical, and even hostile. Some time elapsed before the editor of the *Journal of Ecology* approached Dr Brian Davis of the Toxic Chemicals and Wild Life Section for a 'scientific critique'. In writing one, Davis was able to appraise the book itself and the way in which it had elicited enthusiastic support and outright condemnation. He took 23 of the most striking examples and figures quoted in *Silent spring*, and checked them against the original papers, the great majority of which proved to be 'reliable scientific documents'. The only error found in the British references was an understatement of the number of counties affected by seed-dressing incidents in 1961, namely 44 and not 34 as cited. Davis (1964*b*) concluded that 'the facts quoted are essentially correct even if there has been a very considerable selection of facts, and interpretation of them has sometimes been rather elastic'.

Meanwhile, a statement had been issued to all staff in the Nature Conservancy, following considerable discussion on the Conservancy's committees. Citing the great controversy aroused by *Silent spring*, the statement required them:

to ensure that any discussion or information given on this subject is dealt with in an objective, factual and dispassionate manner. No action should be taken or comment made which can in any way imply any division of opinion within the framework of official scientific organisations.

The statement accompanied a guide to staff, which set out the points emerging from *Silent spring*, the Conservancy's views on them, and the action which had already been taken on toxic chemicals.[13]

The Parliamentary Questions soon began to flow. One sought information on what research was in progress on the 'probable cumulative effects of chemicals which remain toxic for long periods in the soil'. The Conservancy drafted a reply for the Minister for Science, describing how the main research of the Toxic Chemicals and Wild Life Section was devoted to studying the cumulative effects of persistent chemicals in the natural environment, including the soil. It was intended to discover how the chemicals came to be concentrated and distributed, with special attention being paid to predatory species. In drafts of replies to Supplementary Questions, the Conservancy described how valuable information was already emerging from these long-term studies. Even incomplete research could assist in indicating possible hazards which ought to be avoided, at least until they were better understood.[14]

The second Shackelton debate

Lord Shackleton used the publication of the British edition of *Silent spring* as the occasion to move a Motion in the House of Lords, drawing 'attention to the ecological dangers and the destruction of wild life resulting from the widespread use of toxic chemicals in the countryside and in gardens'. There was a pressing need for more research, education, and greater restraint in the use of chemicals. The Motion was debated in March 1963, in conjunction with another, drawing attention 'to the multiple and increasing dangers to health and to life arising from the contamination of food, air and water by toxic chemicals'. The debate lasted five hours.[15]

No longer did anyone apologize for speaking on a trivial subject. The potential impact of toxic chemicals was now of considerable interest because of its implications for the wider human environment. Members of the House of Lords described how the human race was confronted not only by 'an occasional dose of poison which had accidentally got into some article of food, but (by) a persistent and continuous poisoning of the whole human environment'. There were already 2 ppm of DDT in the bodies of Englishmen, and 11 ppm in Americans. In destroying 'the natural environment of living things', man might be destroying himself. Although there was no proven link between the pesticide residues and the increased incidence of cancer and other disorders, it was unwise to assume that such compounds remained inert in the human body for all time. Speakers in the debate recalled how the drug thalidomide had once been regarded as safe. Were human beings coming into contact with other, equally dangerous, substances?

Because of their 'distinctively scientific implications', the first government minister to speak on the Motions was Lord Hailsham, the Minister for Science, who sought to assure the House that there was no cause for general concern over the effects of agricultural chemicals on human health. People were living longer and eating better food. Chemical sprays and pesticides were playing a significant part in 'the extraordinary development of agricultural production and efficiency in this country', and the Medical Research Council had found no evidence of any residues in foodstuffs leading to death or illness. The compounds had, in fact, played an important role in conquering diseases. Lord Hailsham recalled how DDT had been used to reduce malaria and to treat 'the louse-ridden contents of the concentration camps', who could so easily have carried typhus all over Europe after the war.

Lord Hailsham contested the view, put forward earlier in the debate, that 'the human environment should be protected against all chemicals unless they are proved to be non-toxic'. The use of chemicals in agriculture was only one example of man's deployment of 'increasingly powerful substances whose effects are sometimes unexpected, occasionally dramatic and often undesired'. It was impossible to reap the benefits of scientific and technological society without running risks. Instead of expecting government to provide an answer to every problem encountered, the proper course was to watch issues as they arose and to create a proper organization for ensuring that scientific opinion was able to influence policy at a sufficiently early stage.

In this context, Lord Hailsham drew attention to the role of the Research Councils, which were composed 'of independent scientists of a wide range of disciplines independent of the executive Ministries but in close touch with them'. As Minister for Science, he saw his task as one of ensuring that their advice on toxic chemicals was given early, heeded closely, and 'not watered down either by political pressure or by administrative convenience'. This had been the purpose behind the setting up of successive working parties and study groups. Lord Hailsham conceded that the pesticide problem, as it affected wildlife, gave cause for some apprehension, but the Research Councils and Ministry of Agriculture had acted 'fairly promptly in anticipating or stopping abuse'. Legislation and a voluntary notification scheme had followed the Zuckerman inquiries. The recommendations of the (Sanders) Research Study Group were being implemented. The Nature Conservancy had been asked to carry out research on the effects of pesticides on wildlife, and the Minister expected to open its new Monks Wood Experimental Station later in the year.

In winding up the debate, the Joint Parliamentary Secretary to the Ministry of Agriculture challenged both the data and approach of those calling for a more vigorous response from the Government. He denied that there was any scientific evidence for the peregrine falcon and other birds of prey being reduced as a result of pesticide use. Whereas Rachel Carson was a publicist (and caution did not take the publicist far), scientists were 'modest and cautious people, by nature and by training'. They preferred to publish results rather than prophesies, but their 'meticulously checked and advanced findings were as impregnable as anything in science can hope to be'.

This kind of Government response did little to satisfy those members of the House of Lords calling for immediate measures. In the words of Lord Aylesford:

What am I, as a poor, wretched land-owner, to do about it? I can, I suppose, try to restrict the use of chemicals on my own home farm, but I cannot remain for ever in a sort of oasis . . . I cannot stop dead and dying birds from landing on my ground . . . I cannot stop what birds inhabit my home from going outside the boundaries and getting a lethal dose there.

There was nothing of a practical nature that an individual or a small group could do, and 'we therefore have to ask for help from the Government'.

Another speaker, Lord Walston, agreed that a balanced approach to the question of pesticides was required, but this meant that 'as the weight shifts on one side of the scales so, in order to redress the balance, you have to shift more weight to the other side'. When the chemicals were first being introduced, there was every justification for helping 'the initiators, the people who wanted to see these new methods adopted', and who had to contend with 'the normal conservatism of farmers, of food processors and of others'. The situation had changed; there were now 'serious and substantial vested interests' behind the pesticides industry. By being so protective of the industry, the Government had shown little sign of recognizing the profound shift in balance that had taken place.

A few days earlier, Lord Walston had met Dr Mellanby and Dr Moore, and he learned at first hand what preoccupied the Conservancy's mind, and what facilites were needed to develop relevant research on pesticide use.[16] He drew on this in the debate, as he criticized those who claimed that there was no proof that pesticides caused the death of wild birds and animals. In many cases, the assertions amounted to no more than an admission that nobody had discovered whether there was any direct correlation between the chemicals and some disease, symptom, or syndrome. This did not mean that there was nothing to fear. Rather, it demonstrated that much more research needed to be done in order to discover whether fears of a casual link were justified.

One of the principal speakers in the debate was Lord Hurcomb, the Chairman of the Nature Conservancy. He congratulated Miss Carson for writing a book which would be read by millions. By dramatizing the dangers of chemical use, she had forced everyone to look afresh, or for the first time, at the question of the build-up of toxic chemicals in the environment. Unfortunately, 'chemists and most of the people on the other side' still denied the mounting evidence, but this was largely because they worked in blinkers. *Silent spring* had done a major service by emphasizing the value of an ecological perspective. Lord Hurcomb conceded that the Conservancy had been 'slow in getting off the mark' but, in its case, this was due to a lack of manpower and resources.

The President's Science Advisory Committee

There was no chance of the pesticide issue dying down. In June 1963, the long-awaited report of President Kennedy's Science Advisory Committee was published. Although a temperate document, carefully balanced in its assessment of risks versus benefits, it added up to 'a fairly thorough-going vindication of Rachel Carson's *Silent spring* thesis' (Greenberg 1963). Having reviewed both benefits and disadvantages of using pesti-cides, the report set out a series of recommendations aimed at securing 'a more judicious use of pesticides or alternative methods of pest control' (President's Science Advisory Committee 1963).

In a commentary prepared for the Minister, officials in the Ministry of Agriculture drew attention to four of the recommendations contained in the report, namely:

(1) the re-assessment and tightening up of controls on existing chemicals;

(2) making it harder for manufacturers to secure approval for new chemicals;

(3) reducing present tolerances in food, which might curtail or eliminate the use of some persistent chemicals;

(4) reducing the use, with a view to eventually eliminating, the more persistent chemicals.

So far, Britain had 'done rather better in the wild life field than have the Americans' but, the officials warned, America would forge ahead in some respects if the proposals in the report were implemented. The evi-dent determination of the President's committee to end the use of the more toxic and persistent organochlorines could lead to the same goal being sought in Britain.[17]

The Minister for Science asked the Research Councils to assess the report, and advise whether there was scope for collaboration between the two countries in pesticide research. Lord Hailsham was particularly interested in proposals for a pesticide monitoring scheme. The Director-General of the Nature Conservancy described the report as 'very fair and lucid', rightly emphasizing the dangers of the build-up of small quantities of persistent chemicals, the continental scale of the problem, and the fact that pesticides were 'just one aspect of the whole environmental contamination problem'.[18]

As in the consideration of *Silent spring,* it was important to bear in mind the differences between the American and British situations. While the ratio of non-agricultural to agricultural land was much greater in America, proportionally more land was sprayed. This was because of the greater chance of insects destroying entire crops, whereas losses in Britain were comparatively small or could be avoided in other ways. Again in Britain, spraying was almost entirely restricted to edible crops grown in small units, whereas, in America, extensive areas of forest and non-edible crops, like cotton and tobacco, were also sprayed, often from the air. Pes-ticides were used extensively to control mammals in America; they were virtually restricted to use against rats, mice, and voles in Britain (Moore 1964*a*, 1965*c;* Mellanby 1964).[19]

There were also important administrative differences. In Britain, a wide range of disciplines collaborated in determining hazards. No tolerance levels were set; a voluntary Notification Scheme operated. On a visit to the United States in the spring of 1963, Dr Moore found that the manufacturers had to supply information to the Federal Department of Agriculture, which scrutinized products for their effectiveness only. The Food and Drugs Administration decided on legal tolerance levels in foodstuffs. At no stage was the opinion of a biologist formally required, and there was little evidence of his opinions having any effect where they were given informally or obliquely. For these reasons, it was much harder to introduce an ecological approach to pesticide matters in the States.[20]

The (Frazer) Research Committee

In its report of 1961, the (Sanders) Research Study Group had emphasized the need for periodic consultations between those engaged in relevant research, both within the Research Councils and elsewhere. The Agricultural Research Council (ARC) and the Nature Conservancy readily agreed to the holding of informal meetings, but Lord Hailsham believed co-ordination and the allocation of responsibilities should be organized on a more formal basis. In his view, this was the only way of maintaining an adequate drive over the entire field of pesticide research.[21]

It was agreed that the ARC should appoint a Research Committee under the chairmanship of a member of that Council, Professor A.C. Frazer. The 13 other members were scientists occupying senior administrative posts. One was Dr Kenneth Mellanby, the Director of the Monks Wook Experimental Station. The commmittee decided to meet twice a year, to include research done outside the Research Councils in its terms of reference, and to involve further scientists in the work of the Research Committee, through their appointment to working parties set up to investigate particular issues.[22]

Without vigilance, a fragmented approach to pesticide research could easily have occurred. The chemical manufacturees were principally responsible for discovering new pesticides, and trials were conducted by these manufacturers and the Ministry's Plant Pathology Laboratory. The impact of chemicals on soil fertility, bees, and other beneficial insects was studied by university departments and a range of bodies in the Agricultural Research Council. Studies of spray drift were conducted by the National Institute of Agricultural Engineering, and of methods of chemical analysis by the Laboratory of the Government Chemist. The Weed

Research Organisation investigated the use of herbicides in weed control. The Game Research Association studied the effects on game birds and animals, and the Ministry of Agriculture and Fisheries covered the fisheries aspect. Research on the effects of pesticides on wildlife was carried out by the Ministry, the Nature Conservancy, and such voluntary bodies as the British Trust for Ornithology and the Royal Society for the Protection of Birds.

Having visited a number of relevant laboratories in 1960/61, Dr Moore emphasized the need for the closest possible liaison with them. He cited the Jealotts Hill laboratories of Plant Protection Ltd, where the staff of 150, which included 50 graduates, exceeded the entire research staff of the Nature Conservancy. Without their material support, there would never be sufficient resources to devote to the ecological aspects of pesticide use.

In a few cases, joint scientific studies were promoted. In a memorandum to the Minister in October 1963, the Director-General of the Conservancy quoted the way in which Shell Research Ltd was analysing about a thousand samples per year from the Nature Conservancy. A combined study was underway with Plant Protection Ltd on the effects of paraquat and diquat on aquatic habitats, and studies were being carried out on the effects of aerial spraying against the pine looper moth, in conjunction with Fisons Ltd and the Forestry Commission.[23]

The (Frazer) Research Committee decided that its first annual report to the ARC, and therefore the Minister, should give a broad account of the existing situation and an indication of which areas of research might be strengthened.[24] The report was completed in November 1963, and published in March 1964. Because of the striking changes taking place in the use and management of land, the Committee emphasized that the use of pesticides should always be kept under close surveillance. Although there appeared to be no major gaps in research, some aspects needed strengthening so that the scale of effort reflected the potential hazards to human health, domestic livestock, and wildlife. The report emphasized the need for more trained personnel and for greater attention to be paid to the very persistent nature of some pesticides. Much greater priority had to be given to developing more selective, less persistent, and therefore safer pesticides (Agricultural Research Council 1964).

The Research Committee recomended that a higher priority should be given to discovering varieties of crops that were more resistant to pests, and to research on biological methods of controlling pests. Even if dramatic advances were made, it was doubtful, however, whether chemicals would be entirely displaced. Even in the longer term, it seemed likely that the best results would come through a co-ordinated programme that used a range of controls, including chemicals, in the fight against pests and diseases.

A greater priority should also be given to the development of new pesticides to combat the emergent resistant strains of pest. The ideal pesticide was one that was effective against the target and left other species unharmed, but the Research Committee conceded that, even if this was possible, the pesticide would hardly be attractive commercially. It was much more profitable to produce a pesticide that killed a wide range of pests and could, therefore, be manufactured in large quantities. This economic hurdle had to be overcome.

The Research Committee was particularly concerned about the incidence of residues in soils and in the bodies of predatory birds. Crops might pick up residues in soil, and the destruction of soil fauna might lead to a decline in the productivity of the soil. Birds and other vertebrates might eat large quantities of contaminated worms and slugs; and the run-off and seepage of contaminated water might harm aquatic organisms.

In practice, most of the research on the effects of chemicals on wildlife continued to be concentrated in the Ministry of Agriculture and in the Nature Conservancy. Following the introduction of the partial ban on aldrin, dieldrin, and heptachlor as seed-dressings, the 200 Pest Officers in the Ministry were instructed to investigate any 'incident' that might involve agricultural chemicals. The breast muscle (flesh) and liver of any birds or mammals suspected of being involved were to be removed for examination by gas-liquid chromatography. In order to put such incidents in perspective, planned surveys were made of particular parts of the countryside. Small mammals were trapped and birds shot for analysis, and a special study was made of predatory birds.[25]

The adequacy of these arrangements was sometimes criticized but the Ministry refused to introduce anything more elaborate. Criticisms of another sort were voiced by the National Farmers Union. At a conference of June 1962, a spokesman warned that pest officers might be spending too much time protecting pests such as the wood-pigeon, when urgent measures were needed to curb the menace. It was quickly pointed out that far more resources were still being devoted to pest control. After three seasons, the Ministry was able to conclude that voluntary restrictions had succeeded in eliminating the kind of large-scale incidents experienced in 1960/61. On the other hand, the surveys indicated that organochlorine residues were still present 'in a fair proportion of the bodies examined'. In some instances, they had a lethal, or near-lethal, effect.[26]

Each Research Council was asked to keep the (Frazer) Research Committee appraised of its work, and of any constraints on progress.[27] In its submission, the Conservancy described how the work of the Toxic Chemicals and Wild Life Section fell into three parts, namely:

(1) studies of the long-term effects of organochlorine pesticides on plant and animal communities;

(2) studies on the long-term effects of herbicides;

(3) background studies of the farmland habitat.

Throughout 1962, discussions continued on how the Toxic Chemicals and Wild Life Section might develop if further resources became available. Under the agreement negotiated with the Laboratory of the Government Chemist (see page 56), a routine programme of monitoring the extent and persistence of environmental contamination began in 1963. In the following 10 years, over 3300 analyses of wildlife specimens were made. Not only were they crucial to the Section's work, but they provided an important stimulus 'for the development of new analytical approaches' in the Laboratory (Laboratory of the Government Chemist 1974). There were, however, practical difficulties. Because of the unpredictable nature of much of the Section's work, it was never possible to attain an even flow of specimens for analysis. There were also long delays before the results of analyses could be provided, and this precluded their being used in conjunction with on-going feeding trials. It soon became apparent that the Section would have to instal some analytical facilities of its own, particularly for the less routine and more exceptional types of analysis.[28]

The most serious impediment continued to be lack of staff. The absence of a vertebrate ecologist meant that the nucleus of the Section remained incomplete, and it was impossible to embark on studies of the flow of pesticides through food chains and their concentration in predatory species. At a meeting of the Conservancy's Scientific Policy Committee in July 1962, Dr Mellanby spoke of the Section's equally desperate lack of scientific assistants. In his words, the Section's scientists could plan their work, and devise admirable programmes, but they had no means of carrying out the extensive surveys and trials that their research required. The situation was most serious in the case of the Head of the Section, Dr Moore, who had to devote so much time to committee and other administrative work. Unless he was given reliable and adequate assistance, his scientific efforts would be largely wasted.[29]

In the Conservancy's Estimate for 1963/64, there was room for only one additonal post of scientific officer (that of the vertebrate ecologist) and two scientific assistants in the Toxic Chemicals and Wild Life Section. The Director-General warned the Office of the Minister for Science that the Conservancy had 'gone to the limit in prejudicing other programmes' in order to give maximum resources to the Section. Nothing further could be done without the Conservancy being given a special mandate, and therefore resources on a commensurate scale, in order to meet the 'emergency situation' that was developing in respect to pesticides and wildlife. To that end, and in addition to the posts included in the Estimates, Mr Nicholson applied for an extra 17 posts for the Section. In January 1963, he learned that the Treasury had agreed to 12 of the posts in addition to those in the Estimates. The additional posts comprised four in the scientific officer class, two in the experimental officer class, five scientific assistants, and one shorthand typist.[30]

By the middle of 1964, the posts had been filled. Ian Prestt left his post as Deputy Regional Officer of the Conservancy's North Region to become the Section's vertebrate ecologist. Dr F. Moriarty and Dr D.J. Jefferies became responsible for toxicological studies on insects and vertebrates respectively; Dr J.P. Dempster embarked on his fundamental field study of the effects of DDT on interspecific relationships. Dr M.D. Hooper was appointed to carry out genetic studies on the impact of toxic chemicals, but soon began an intensive study of hedgerows and their changing distribution. A student from the second year of the Conservation Course at University College, London, E. Pollard, was appointed to study the ecology of the insect fauna of hawthorn hedges.

Later staff changes may be summarized here. C.H. Walker resigned in 1967; and his assistant, M.C. French, succeeded him as the Section's analyst. R. Crompton resigned in 1965. J.L.F. Parslow joined the Section as its information officer in 1967, and soon became closely involved in studies of the effects of toxic chemicals on marine birdlife. Dr A.S. Cooke was appointed in 1968, and two former members of the Infestation Control Laboratory, Dr R.K. Murton and N. Westwood, joined in 1970. The number of scientists and scientific assistants in the Toxic Chemicals and Wild Life Section never exceeded 26 at any one time.

The official opening of Monks Wood

For many months, embryonic Sections of the Conservancy's new experimental station had been operating from temporary premises.[31] The Toxic Chemicals and Wild Life Section shared accommodation with the Woodlands Research Section at St Ives. The Conservation Research Section moved to the Monks Wood site in October 1961, and shared a First World War army hut with the Station's Director, Dr Kenneth Mellanby, who had obtained it for £200! *Nature*, described Dr Mellanby as 'one of the few senior entomologists with experience in both medical and agricultural entomology'. He had been the first Principal of the University of Ibadan in Nigeria, and had, since 1955, been head of the Department of Entomology at Rothamsted (Anon. 1961).

Because of the ceiling of £100 000 imposed by the Treasury, only the first phase of development could be carried out. The accommodation intended for the Vertebrate Ecology Section was left out, and the space allocated for the three other Sections was substantially reduced. Although the Treasury had approved the acceptance of the lowest tender for erecting the buildings in March 1962, work was severely delayed. Only £35 000 were allocated for the financial year, 1962–63. Because this represented only about six months' building time, construction could not begin until halfway through the financial year.

One of the first actions of Dr Mellanby was to protest that this would damage staff morale. The Conservancy had recruited a team of first-class scientists, and they had every reason to feel let down by the further delay in providing them with adequate and permanent accommodation. It was not until May that the Conservancy persuaded the Treasury to allow building to start. The building contractors moved onto the site in June, but further delays were caused by the winter of 1962–63, which was one of the most severe on record. All building came to a standstill for two months. The Toxic Chemicals and Wild Life Section was the last to move onto the site of the new Station in mid-September 1963.[32]

Monks Wood became the largest of the Conservancy's stations, and was officially opened by Lord Hailsham on 28 October 1963.[33] Local newspapers described how the Lord President of the Council and Minister for Science, who had so recently been a contender for the office of Prime Minister, had defied doctor's orders to rest his injured foot, and had made a special car journey to Monks Wood to open the new station. Drawing on his speech in the House of Lords earlier in the year, Lord Hailsham (1963) directed most of his opening address to the effects of toxic substances on wildlife. He described how the question affected several government departments, and had aroused 'strong emotions'. In urging a balanced and cautious approach to the resolution of the problem, he emphasized that most of the damage to wildlife had arisen from the indiscriminate use of a particular group of chemicals. Now that the dangers had been recognized, effective controls had been introduced. The episode demonstrated not only the dangers involved in the use of these substances in the countryside, but also the effectiveness of the defences that had been erected by the Government against their misuse.

THE ORGANOCHLORINE REVIEW

In Britain, the role of the Inter-departmental Advisory Committee on Poisonous Substances continued to be crucial in the regulation of pesticide use. The Conservancy's representative on the Committee was still Mr R.E. Boote, who described in late 1962 how the Conservancy often took the lead in challenging the safety aspects of a pesticide, only to discover that other members of the Committee had similar misgivings. One question led to another, and significant changes were sometimes made while consideration was being given to products submitted for the approval of the Committee. After the events of early 1963, the Advisory Committee seemed even readier to take a strong line in regulating the introduction of pesticides. It was now assumed that manufacturers should have the necessary staff and resources to assess the impact of their products on wildlife.[1]

Despite these encouraging trends, the Conservancy believed the Advisory Committee should become even more objective and scientific in its approach. The time had come to introduce a licensing system, administered by an independent statutory body and free from any vested interests in the marketing and use of pesticides. Mr Boote spoke of how, 'once we have more scientific and technical knowledge, notably effective field trials and tests in our possession, we should be able to press forward to the inevitable full legislative control of all toxic chemicals'.

There were, however, signs that this would be an up hill struggle. Not only might the Advisory Committee drag its feet on key issues, but there were serious deficiencies in the application of the voluntary Code of Notification; certain major uses of chemicals were not covered. After a number of unsuccessful attempts to interest the Ministry of Agriculture in the matter, Mr Boote drafted a minute in November 1962, which described the need for the Conservancy to take the initiative in bringing, for example, the highway authorities, water boards, drainage authorities, and the British Transport Commission, under the existing, or a new, Code of Notification.[2]

'Chemicals for the gardener'

Resolutions from the Royal Society for the Protection of Birds and from the International Committee for Bird Protection in 1963 drew attention to another deficiency in the voluntary scheme, namely to its failure to regulate the use of toxic chemicals by the amateur gardener. In response, the Ministry of Agriculture argued that there was insufficient evidence of any harmful effects to warrant a ban being imposed on the use of aldrin, dieldrin, and heptachlor in gardens. The Conservancy sympathized with this viewpoint, but argued that every opportunity should be taken to reduce the number of chemicals available to the public, particularly until more was known about the behaviour of these pesticides. At the suggestion of the Conservancy's Committee for England, the Royal Horticultural Society was asked to assist in publicizing the dangers of the indiscriminate use of pesticides in the garden. In May 1963, the Conservancy wrote again to the Ministry, expressing its concern over the use of toxic chemicals in horticulture and gardening.[3]

Matters came to a head in June 1963, when the Ministry published an official guide to gardeners on the proper use of chemicals, called *Chemicals for the gardener.* The use of organochlorines was advocated, despite a preface to the guide which said that all preparations had 'been considered for possible dangers to humans, domestic animals and wild life'. As far as was known, none of the garden packs, when used according to the directions, was likely to be harmful to birds or other wildlife in the garden. Not only was the Conservancy angry that this ignored strong scientific evidence that organochlorines were harmful to wildlife in some situations, but it was particularly annoyed that the Conservancy had not been consulted on this point before the publication of the guide. The senior officers of the Conservancy first saw the booklet a week after publication. To make matters worse, the guide had been published at the start of the first National Nature Week ever to be organized, when the media was particularly alert for any instances of conflict in the Government's policy on nature conservation (Ministry of Agriculture 1963).

These anomalies were brought to the attention of the Minister for Science and, at his request, the Director-General prepared a draft which recalled that particularly since the Lords' Debate on the 'Dangers of toxic chemicals' in March 1963, attempts had been made to ensure that all official pronouncements reflected a common approach. This was especially important in view of scientific and technical complexities, the rate at which fresh evidence was becoming available, and the strong feelings aroused by the use of pesticides. Such inter-departmental liaison, he wrote, required that the appropriate research council, the Nature Conservancy, should be consulted. A few days later, Lord Hailsham wrote to the Minister of Agriculture. In his letter, he stressed that 'my main concern is that we should not say anything which implies that danger to wild life can be disregarded unless this has been clearly established by scientific evidence and also that we should take special care to show continued willingness to review new evidence and to act vigorously if the situation so requires'.[4]

Matters were going from bad to worse. In the House of Lords, Lord Shackleton put down a Parliamentary Question asking whether the booklet represented the wise use of toxic chemicals, and whether 'the Government's advisers on wild life were consulted before its issue'. The Parliamentary Secretary to the Ministry of Agriculture replied in the affirmative, adding that:

the advisory booklet was published under the Agricultural Chemicals Approval Scheme to help gardeners choose the right chemical by listing those which have been found effective in controlling troublesome pests, diseases and weeds . . . No product is approved unless the chemical it contains has also been considered and cleared under the notification scheme from the safety point of view.

Because the Inter-departmental Advisory Committee included a representative of the Nature Conservancy, by inference the Conservancy had been consulted.[5]

Meanwhile, the RSPB had responded by publishing a leaflet warning the public not to turn their gardens into 'a death-trap for birds'. Even comparatively safe compounds should not be used except in an emergency. The public were urged to write to their members of parliament, pressing for the withdrawal of organochlorines and for the publication of lists of insecticides that gave adequate warning of the dangers to wildlife.[6]

There was no possibility of the issue dying down. Immediately after the Whitsun Recess, the Parliamentary Secretary apologized to the House of Lords for claiming, in his earlier statement, that the Advisory Committee had seen and approved the booklet prior to its publication. He had subsequently found this to be untrue. Two days later, Lord Shackleton put down a further Parliamentary Question asking the Minister to 'urgently consult the Government's advisers on wild life and in the light of their advice (to consider) whether this publication should now be withdrawn'.[7]

On the same day, the Minister of Agriculture replied to three questions on the same subject in the House of Commons. He said that the Advisory Committee had only imposed restrictions on the use of pesticides as seed-dressings and that because the booklet did not cover this type of use, there had been no need to refer it to the Committee. The Minister acknowledged that there was strong feeling among certain societies that the restrictions should be extended, and he was therefore asking the Advisory Committee to carry out a further examination of the question.[8]

Although this statement ensured that the use of pesticides in the garden and elsewhere would be looked at afresh, it still left an important point of principle unresolved. The publication of the booklet had brought to a head the question of the status of the Nature Conservancy, a question that had first arisen during correspondence between the Conservancy and the Minister for Science. Despite the strongest and most urgent appeals for the withdrawal of aldrin, dieldrin, and heptachlor, and a reduction in the use of DDT and gamma-BHC wherever less persistent compounds could be used, the Minister had rejected the Conservancy's advice on the grounds that the question was being reviewed by the Advisory Committee, on which the Conservancy sat.[9] The Conservancy regarded the reply as totally unsatisfactory because it overlooked the fact that it was the Conservancy, and not the Ministry of Agriculture or some *ad hoc* committee, that had the statutory responsibility for protecting wildlife. It was not enough to refer questions on pesticides in relation to wildlife to the Advisory Committee on the premise that a representative of the Conservancy sat on that Committee. This did not ensure adequate consultation or recognition of the status of the Conservancy.

It was against this background of unease that some members of the House of Lords, including Lords Shackleton and Hurcomb, heard the minister's statement, as relayed to them by the Parliamentary Secretary. They were joined by Lord Morrison of Lambeth who, as Lord President of the Council in the late 1940s, had played the principal role in establishing the Conservancy. With other members of the House, Morrison insisted that the Conservancy should be given every opportunity to express its views to the Minister and the Advisory Committee.[10]

The Conservancy was more than ever determined to press its case and, in a fresh memorandum to the Advisory Committee, it reiterated its earlier demands. The latest information was included on the decline of certain bird species and on the presence of pesticide residues in corpses found in the field. Almost invariably, the highest levels of residue were found in birds of prey. Although far too little was known about the significance of residue levels, the striking difference between those levels recorded in species at the top of the food chain and the remainder was surely significant. The notification of new compounds should be preceded by much more testing in the laboratory and field. Aldrin, dieldrin, and heptachlor should be completely withdrawn, and the less toxic DDT and gamma-BHC should be discontinued wherever possible. A monitoring service should be established.[11]

As expected, there was strenuous opposition to the proposals from those who argued that no further restrictions should be introduced until the dangers were proved beyond doubt. Dr Moore warned that several species could well be extinct before that happened and that there could be no delay or compromise over the banning of the use of the compounds in sheep-dips, domestic gardens, and on government-controlled land. In an internal minute, Dr Moore argued that the Conservancy should only agree to the use of the compounds in one or two localized and specialized instances where no real substitute was available.[12]

The Conservancy failed in its attempt to persuade the Advisory Committee to introduce an immediate ban. It was argued that this would be premature because only some aspects of the pesticide problem had been examined. Instead, the Advisory Committee decided to set up a formal review in order to establish causal relationships, and to take account of the views of the pesticide trade and users with respect to the British and American data on wildlife losses. Although the decision represented a further delay in taking action and was tantamount to an admission that the normal procedures of the Committee were inadequate, the Conservancy hoped the continued uncertainty would prove so troublesome to the trade that some firms would stop producing the offending pesticides, whatever the outcome of the Committee's deliberations. On the wider question of the status of the Conservancy on matters affecting wildlife, little real progress was made either then or at a later date. It was one of the crosses the Conservancy had to bear throughout its life.

The peregrine falcon

Under the chairmanship of Sir James Cook, the Advisory Committee began its review of organochlorine pesticides in the autumn of 1963. The Conservancy set about revising its memorandum to the Advisory Committee in order 'to consolidate every atom of evidence and to make the case for immediate action before next Spring as watertight as possible'. In his fresh submission, Dr Moore argued that four questions had to be answered before any further recommendations on the control of pesticides could be put forward.[13] These were:

1. Have any wildlife species declined since the introduction of organochlorine insecticides?
2. Has the reproductive capacity of any species decreased in that period?
3. Are organochlorine insecticide residues widely distributed in bird and mammal populations?
4. Are the amounts found large enough to support the hypothesis that declines in numbers and reproductive capacity are due to lethal and/or sub-lethal poisoning by organochlorine insecticides?

Until recently, most of the Conservancy's scientific data had come from scientists outside the Conservancy. Members of the Infestation Control Laboratory of the Ministry of Agriculture had been particularly helpful in this respect. At long last, the Conservancy's own scientists were now in a position to give advice, based on their own findings. In his submission to the Advisory Committee, Dr Moore cited the interim results of a wide variety of surveys and analyses being caried out by his Section.

In order to understand the rate at which this information flowed from the incipient programmes of the Toxic Chemicals and Wild Life Section, reference must be made to the remarkable headstart resulting from the working relationships that had evolved over the years between personnel of the Conservancy and the voluntary bodies. In a paper published in the annual *Handbook of the Society for the Promotion of Nature Reserves,* the Director-General of the Conservancy described how there were groups of talented people working through innumerable organizations, each seeking 'a new and more harmonious relationship between man and nature'. One of the most recent and encouraging trends had been the breaking down of hard and fast divisions between these various groups as they realized that each one had an indispensible part to play in seeking improvements over an ever-broadening range of environmental concern (Nicholson 1965).

This close collaboration was amply demonstrated in the field of bird protection. It had been the policy of the Nature Conservancy not to appoint any staff to work explicitly on birds. In evidence to the Commons Select Committee on Estimates in 1958, Mr Nicholson described how it was much more economical to make grants to universities and such bodies as the British Trust for Ornithology (BTO). Mr Nicholson had himself played a key role in organizing and developing field ornithology on a broadly ecological basis. He had been a co-founder of both the BTO and Edward Grey Institute at Oxford, and had served as secretary and chairman of the latter body. Rather than embark on its own programmes, Mr Nicholson believed the Conservancy should stimulate and strengthen the activity of these voluntary bodies. It already provided considerable financial assistance in the form of commissions.

Indirectly, the Conservancy came to be involved in launching several major research and conservation measures. The Common Bird Census of the BTO was started at the Conservancy's request in 1961, with the aim of comparing, on an annual basis, breeding populations over as large an area as possible. Previously, ornithologists had only been able to express an opinion as to whether populations had been rising or falling—now they could be rather more precise. Following the severe winter of 1962–63, the Census indicated, for example, that the wren (*Troglodytes troglodytes*) population had fallen by 80 per cent. Such data were to prove invaluable for assessing the effects of toxic chemicals on birdlife generally.

Although appointed to work on other aspects of natural history and nature conservation, the staff of the Conservancy played a key role in ornithological research and bird protection—by pursuing these activities out of working hours. In some cases, the studies became so important to the Conservancy, as well as to the wider conservation movement, that dispensations were granted for them to be pursued or completed in working time. Norman Moore (1957) published an account of the past and present status of the buzzard (*Buteo buteo*) in the British Isles. Derek Ratcliffe had been employed as a botanist, but he devoted much of his spare time to the study of the peregrine falcon (*Falco peregrinus*) and other predators. These studies, extending over the period 1945 to 1961, formed the basis of a paper which he published in *Ibis* in 1962 on the breeding density of the peregrine and raven (*Corvus corax*) (Ratcliffe 1962).

These extra-curricula activities, both on committees and in the field, greatly strengthened the Conservancy's role in the conservation movement by providing staff with a far greater range of insights and experiences than would otherwise have been possible. It made it easier for the Conservancy to discern significant links between what might otherwise have appeared disparate and unimportant trends in the natural environment. An example of this can be cited. In 1958, Dr Ratcliffe published a note in *British Birds*, drawing attention to the increasing frequency with which broken eggs were found in peregrine eyries. Whereas there had been scarcely any references to breakages before about 1950, Ratcliffe found 13 instances between 1951 and 1956, representing over a fifth of the nests visited. In most cases, if not all, the eggs seemed to have been deliberately smashed and eaten by one or other of the parents, but the reasons for such abnormal behaviour were not known. Because of the birds' catholic choice of prey, it was unlikely to be a response to food shortages. Ratcliffe hoped that his note would encourage other ornithologists to record similar incidents. Within a few years, knowledge of such incidents, and of their frequency, was to arouse considerable interest among those studying the effects of pesticides on wildlife (Ratcliffe 1958).

Signs of a significant decline in population of some bird species were noted by Richard Fitter, on behalf of the Council for Nature, after the breeding season of 1960. From information gathered from observers in a number of counties, the kestrel (*Falco tinnunculus*) and sparrowhawk (*Accipter nisus*) appeared to be principally affected. The Nest Recording Scheme of the BTO also suggested a decline of the sparrowhawk in the south and east of England since 1960 (Cramp 1963). There was also concern for the peregrine falcon (*Falco peregrinus*) in Cornwall. The reasons for the decline were unknown, but R.B. Treleaven drew attention to the apparent excess of females in the population, and the possibility of changes in the amount of food available and degree of human interference (Treleaven 1961).

Particularly valuable information could be derived from surveys of bird populations undertaken, or in progress, before organochlorine pesticides were used on a large scale. These provided a base line from which subsequent changes could be measured. The peregrine falcon was one of

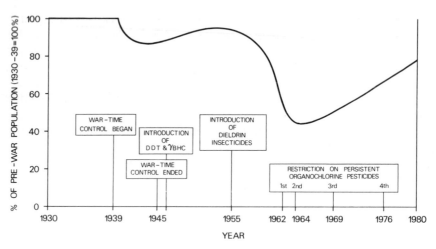

Fig. 7.1 Decline and recovery of the peregrine falcon population.

the species to have been surveyed in this way (Fig. 7.1). It was the largest falcon indigenous to Britain and had a comparatively small and stable population. It was the perfect raptor, majestic in appearance, unrivalled in speed of wing, and almost worldwide in its distribution. The prize of falconers from medieval times, the bird continued to excite the admiration of all who knew it (Lowery 1969).

The opportunity for a further survey of the breeding population of the peregrine arose in September 1959, when the Home Office sought the advice of the Nature Conservancy as to whether special protection given to the bird under the Protection of Birds Act 1954 should be removed because of complaints that the peregrine was inflicting heavy losses on racing pigeons. No one could deny that the peregrine fed on live birds or that its principal prey, south of the Scottish Highlands, was the domestic pigeon, but the Conservancy was extremely sceptical of allegations that the birds were taking 100 000 homing-pigeons a year, at an estimated cost of £500 000 to pigeon-fanciers. There were even claims that the pigeon-fancier himself might become extinct! Whatever the validity of these claims, the Conservancy insisted that no change in the protection accorded to the peregrine should be contemplated until an up-to-date and objective assessment had been made. The Home Office agreed to the Conservancy's suggestion that a survey should be mounted of the peregrine's breeding population in the whole of the British Isles before a formal decision was taken. It was calculated that three years would be needed: the first for exploratory work and for setting up a network of recorders, the second for the survey itself, and a third for checking the results. Realizing that there would be no relaxation in protection without a survey being carried out first, the organizations representing pigeon-fanciers agreed to co-operate by providing details of pigeon losses, flight routes, and the dates of races.[14]

The British Trust for Ornithology (BTO) was awarded a contract worth £2 212 by the Conservancy to establish the number and distribution of the breeding population of peregrines and, in so far as it was possible, to investigate such questions as winter distribution, non-breeding stock, immigration, and emigration. Dr Derek Ratcliffe was appointed in a personal capacity by the BTO as the main organizer. In order to emphasize the impartiality of the survey, it was a strict condition that all individual records gathered by surveyors would remain confidential to the BTO, and would not even be divulged to the Conservancy. In the event, it was too late to attempt much in 1960. Most potential helpers were engaged in other work and, despite his appointment in a personal capacity, it soon became clear that Dr Ratcliffe would have to devote some of his official time to the survey during the breeding season if adequate progress were to be made. Because of the imperative need to complete the Conservancy's report on the Highland Vegetation Survey, a dispensation could not be granted until the 1961 season.[15]

Ratcliffe (1980) has recently discussed the results of the survey and their far-reaching repercussions. Although the census of 1961 was not as complete as had been hoped, enough was learnt to indicate that, far from becoming more abundant, the peregrine had experienced a dramatic and abnormal decline in population since the previous survey of 1948–49. Many territories in southern England and Wales were deserted, and few pairs had tried to nest. In northern England and southern Scotland, most territories were still occupied but many pairs had failed to nest and few young were reared. Despite its obvious limitations, the evidence was strong enough to convince the Home Office that maximum legal protection for the peregrine falcon should be maintained indefinitely. The survey had also shown the great role of amateur observers in population and mortality studies, when organized by a body such as the BTO.[16]

The objective of the peregrine survey soon changed to one of an urgent search for ways of 'conserving a dwindling if not disappearing species'. Dr Ratcliffe has described the search as 'a case of detective work with few clues to follow at first'. It was important not to focus public attention on the possibility of toxic chemicals being the cause of the decline until 'at least good circumstantial evidence' was available. Plans to publish a report on the 1961 survey were abandoned by the BTO until more information was available and, in the event, a paper on the changing status of the peregrine was not published until 1963 (Ratcliffe 1963).

Norman Moore had known Derek Ratcliffe for many years, and had kept in close contact throughout the survey. Both had already come to the conclusion that pesticides were the most likely reason for the decline when, in May 1961, the Director-General asked the Toxic Chemicals and Wild Life Section to investigate the possibilities of a link more closely.[17] There was no evidence to suggest that the peregrine was suffering from the same kind of food shortages that had afflicted the buzzard following the outbreak of myxomatosis among rabbits. It seemed much more likely that the peregrine was affected in the same way as foxes that had died after eating contaminated pigeons. Dr Moore suggested comparing the distribution of deserted and successful peregrine nests with the known use of chemical pesticides.

The picture became clearer, and even more disturbing, when the results of the 1962 census became available. Until the late 1950s, historical records suggested that the population had been unusually stable for a British breeding species. It showed little response to climate or other natural changes. The breeding strength remained fairly steady at about 650 pairs between 1930 and 1939. Although numbers had been deliberately reduced during the war in an attempt to protect carrier-pigeons from predation, the population soon recovered. Evidence for a decline in population appeared first in southern England. The 1961 population was two-fifths the pre-war one. Only 82 pairs (19 per cent of the territories visited) had reared any young. In 1962, only half the known territories were occupied, and the number of successful nestings had fallen to 68 (13 per cent). The peregrine had virtually disappeared as a breeding species in southern England and Wales, and had been reduced to very low levels in northern England and southern Scotland. Even parts of the Highlands had been affected.

In his paper in *Bird Study*, Ratcliffe (1963) reviewed the alternative causes for decline. Because of the cosmopolitan nature of the species, and the fact that it was suffering a serious and synchronous decline in other parts of the northern hemisphere, it seemed impossible to relate the decline to any change in climate or reduction in food supply that would be sufficiently ubiquitous and significant. There was likewise no evidence to suggest that the decline was caused by the outbreak of diseases or unprecedented human persecution. Instead, it seemed that the peregrine was the secondary victim of 'agricultural toxic chemicals, through repeatedly taking prey which carried sub-lethal doses and so building up poison in the body, finally to a fatal concentration'. From American work, it was known that levels of 3.3 milligrams of aldrin per kilogram of body weight, or 5.8 milligrams of dieldrin, could be lethal to adult quail (*Colinus virginianus*), and that the chances of quail eggs hatching were reduced significantly when the adult birds had been fed during winter on a diet containing 1.0 ppm of dieldrin (de Witt 1956).

The wave-like, northward spread in the decline of peregrine numbers reflected the manner of introduction of DDT and, later, dieldrin, aldrin, heptachlor, and gamma-BHC. These were first used extensively in arable areas of lowland England. At first, observers were puzzled by the decline in the Scottish uplands where pesticides were thought not to be used. They were soon disabused. In December 1961, Ratcliffe received a letter from a farmer and keen ornithologist in Argyllshire, who wrote:

I think you are very much mistaken in thinking that seed-dressings are limited. I very much regret to say that we found dead birds (Pigeons, rooks, gulls and a few small birds) after sowing our rape and kale in June this year, and when I looked into things the dressings were 'Dieldrin'. We sowed 35 acres. All Kintyre and Islay have very extensive arable farms . . . Also Ayrshire, Galloway and Kirkudbright-shire are full of cropping farms and dressing of seed seems to be automatic on the part of seedsmen.

A former warden on the Beinn Eighe National Nature Reserve reported pheasants and jackdaws having died almost certainly from toxic chemicals in East Ross. This still left some areas where casualties were recorded and yet, patently, no chemicals were being used. One such puzzling area was St Kilda. In January 1962, Ratcliffe wrote, 'I am beginning to think that Peregrines in these remote districts move out in winter, due to dearth of food, and may reach districts where they can take poisoned prey'.[18]

In view of the weight of circumstantial evidence, considerable interest was attached to chemical analyses of the peregrines' eggs. In July 1961, Dr Ratcliffe flushed a peregrine from an eyrie where she was incubating two obviously addled eggs. The nest was located in a hill glen in Glen Almond, Perthshire, which was used entirely as sheep walk and grouse moor. It was, however, within easy reach of the fertile districts of Strathearn and Strathmore (Moore and Ratcliffe 1962). One egg was broken, but the other was sent to the Laboratory of the Government Chemist for analysis, as soon as arrangements for the use of the Laboratory by the Nature Conservancy had been made (see p. 56). By that time, the egg was a year old! In a report of September 1962, Dr H. Egan of the Laboratory reported that the whole sample of cracked and decayed egg, including the shell, was found to contain, uncorrected for recovery, 115 micrograms of pp'-DDE, 50 micrograms of dieldrin, and 28 micrograms of heptachlor-epoxide. In acknowledging the report, Dr Moore commented that it gave 'very firm support to our hypothesis that the decline of the Peregrine Falcon is due to toxic chemicals; until recently this hypothesis has been considered very far fetched by many interested in the problem'.[19]

Prospects for the peregrine appeared bleak. The bird seemed defenceless, for there were no signs of any physiological adaptation to the poisons. By 1962/63, the only districts where the population had remained stable were those inland parts of the Highlands and Islands of Scotland where little or no pesticides were used, and where the food supply was plentiful throughout the year, thereby making it unnecessary for the birds to move elsewhere. Even here, the peregrine was not entirely out of danger because there was nothing to stop contaminated birds coming into the districts and becoming prey to the peregrines. Dr Ratcliffe believed the situation was likely to get worse over the country as a whole because of the time-lag in the appearance of cumulative effects. Numbers and breeding success would soon be reduced to an extremely low level, 'and final extinction could not be dismissed as a possibility' (Ratcliffe 1963).

Alarmed at the accumulating evidence, the BTO and RSPB organized a conference in March 1963 in order to focus attention on the position in the light of modern knowledge of the birds' relationship to their environment and prey species. Attendance at the conference in Cambridge was by invitation only, and included not only naturalists and falconers but also landowners, sportsmen, gamekeepers, pigeon-fanciers and others whose interests sometimes brought them into conflict with birds of prey. Papers were given on the current status of individual predator species, and attempts were made to identify the causes of population decline. The following resolution was passed without dissent:

This Birds of Prey Conference of sportsmen, falconers, pigeon fanciers, land-owners, farmers, gamekeepers, naturalists, research scientists and others, finds conclusive evidence of an alarming decline in numbers of birds of prey in Britain over the past six years.

The Conference attributed the decline largely to the use of toxic chemicals, and urgently recommended that:

the agricultural, horticultural and forestry use of such chemicals especially persistent chlorinated hydrocarbons, should be critically re-examined and where necessary reduced (Cornwallis 1963).

Residues and other bird species

The peregrine falcon had become a *cause célèbre*. It provided dramatic evidence of a decline in species population and of reproductive capacity since the introduction of organochlorine pesticides. How far were the fortunes of the peregrine shared by other species? It was known, for example, that some species had also experienced a population decline but, in the case of the wryneck (*Jynx torquata*) and the red-backed shrike

(*Lanius collurio*), this had started long before the introduction of insecticides. In others, the evidence was more suggestive of a causal link.

In August 1963, Derek Ratcliffe and J.D. Lockie informed the
Director-General of the Conservancy of a paradoxical situation that was
developing in respect to the golden eagle (*Aquila chrysaetos*). There
appeared to be at least 400 breeding pairs compared with a previous
recorded maximum of 290. Sheep farmers and crofters were becoming
increasingly vociferous in their calls for the removal of legal protection
from the species. At the same time, evidence was accumulating of a
serious decline in breeding success, caused by non-breeding and by
broken or addled eggs, particularly in the west Highlands. Analyses of
eight eggs taken from different eyries had revealed the residues of four or
five organochlorines, some in 'disturbingly large' quantities. The final
results were not yet available, but Ratcliffe and Lockie warned that the
whole situation appeared identical to that of the peregrine, except that
'we have come into it at an earlier stage and the picture is not so clear cut
as in the Peregrine since the main probable source (but not the only
source) of the pesticides is sheep which are not the staple diet of eagles
throughout the year in all areas'.[20]

Ratcliffe and Lockie were well placed to assess the significance of the
changes recorded in their sample study of a wide area of the western
Highlands during 1963. A series of detailed investigations of the status
and feeding habits of the golden eagle had been made during the 1950s
(Lockie and Stephen 1959).[21] The new survey revealed that the proportion of pairs rearing young successfully had fallen from a level of 72 per
cent, as recorded between 1937 and 1960, to one as low as 29 per cent
from 1961 to 1963. There had been a marked increase in reports of
abnormal behaviour and of broken eggs since 1960. There were cases of
eyries being repaired and of brooding taking place, despite there being
no lining to the nest or eggs laid.

Lockie and Ratcliffe (1964) attributed these sudden changes to the
presence of residues, and particularly to dieldrin, in both the adults and
eggs. There was no evidence of any increase in shooting, trapping, deliberate poisoning, egg-collecting, or even casual disturbance to account
for the scale of the decline. Neither had there been any dramatic change
in climate or in food supply. More positively, dieldrin, gamma-BHC, and
DDE were found in ten eggs taken from seven eyries widely spread over
Argyllshire, west Inverness-shire, Ross-shire, and Sutherland in 1963.
Most of this was sheep walk, with little or no cultivated land. The only
possible source of contamination was the large number of sheep-dips.
Sheep carrion made up a large part of the eagle's diet. Significantly, there
had been no change in the breeding success of the eagles of the eastern
Highlands, where there were fewer sheep and the birds seldom fed on

dead sheep. In those parts, eagles depended much more on grouse and blue hares, and took red deer as carrion.

DDT had been widely used in sheep dips since the late 1940s. Gamma-BHC was first used in the early 1950s, and dieldrin had been used since the late 1950s. The eagles probably ingested the pesticides from both skin and fleece, and from the fat and flesh of the sheep. There was evidence for organochlorines being able to pass directly through the skin. A sample of sub-cutaneous fat from a sheep in West Ross was found to contain 0.6 ppm of dieldrin.

At least two-thirds of the British breeding population of eagles lived in this type of sheep country. Lockie and Ratcliffe (1964) warned that the poor breeding-success would lead to a situation where there were no new birds to replace the existing adult population. If the golden eagle followed the same course as the peregrine, adults were likely to die prematurely as pesticides accumulated in their bodies.

No area was free from contamination. In evidence to the Advisory Committee, the Nature Conservancy gave the results of analyses of 16 mammals, 65 birds, and 18 eggs, taken from a variety of sources, and carried out by the Toxic Chemicals and Wild Life Section using paper chromatography, and by the Government Chemist using both gas and paper chromatography. Some of the analyses had also been done by the Department of Agriculture for Scotland. High residues were found not only in those species which had sustained a marked fall in population, but even in specimens from nature reserves. Dieldrin and DDT, and its metabolites, were found in two eggs of the little tern on the Scolt Head National Nature Reserve in Norfolk. An egg from the heronry on the High Halstow National Nature Reserve in Kent contained 27 ppm of dieldrin and 30 ppm of DDE. From a conservation point of view, a level of 15 ppm of dieldrin in the liver of a red kite (*Milvus milvus*) from Wales was particularly disturbing.

In a paper published in *Nature* in March 1964, Moore and Walker (1964) gave the results of analyses carried out on 85 bird corpses and 72 eggs collected in the field, generally in the absence of circumstances suggestive of poisoning. Residues were found in birds and eggs from 37 counties and from 20 of the 21 taxonomic families examined, living in terrestrial, freshwater, and marine habitats. Following a visit to the United States and the chance to see work at first hand, Dr Moore became more than ever convinced that a systematic search would lead to the discovery of residues far away from sites where the pesticides were used.[22] During 1963–64, high priority was given to analyzing further examples of marine life. Ninety sea-birds' eggs were collected from Scolt Head, the Farne Islands, St Abbs Head in Berwickshire, and Great Saltee Island in County Wexford. All contained dieldrin and pp′-DDE; 70

specimens had residues of pp'-DDT and/or its metabolite pp'-TDE; 47 eggs had traces of gamma-BHC (Moore and Tatton 1965).

On land and sea, the surveys provided further confirmation that the highest residue levels were likely to occur in raptorial and fish-feeding species (Fig. 7.2). Much higher levels were recorded, for example, in the eggs of the peregrine falcon, great crested grebe and heron, than in the herbivorous Canada goose, the omnivorous pheasant, or the carrion crow (Moore and Walker 1964). During 1963, 61 eggs were analysed from 64 different nests as part of a survey of the levels of pesticide in raptors and corvids in northern England and Scotland. All contained organochlorine residues, but the amounts found in the raptors' eggs tended to be much higher (Ratcliffe 1965a).

Never before had the evidence and its implications been presented so clearly and forcefully; the proof that residues were to be found in fresh-water and marine ecosystems represented a major breakthrough for the Toxic Chemicals and Wild Life Section. Because the compounds were rarely used in these ecosystems, they could only have reached bodies of water by one or more of four routes, namely, in the run-off from treated land, from aerial drift, from contact with the bodies of migratory species, or from the dumping of containers after use. Dr Moore created quite a

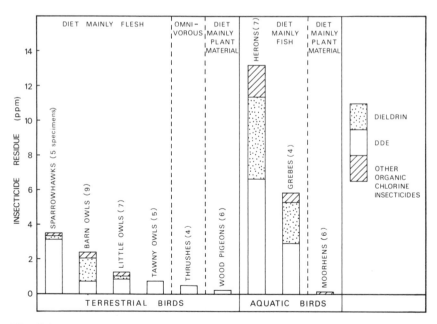

Fig. 7.2 Average concentration of organochlorine insecticide residues in breast muscle of 9 bird species.

stir when he cited these findings at a symposium on 'Ecology and the industrial society', convened by the British Ecological Society in April 1964 (Moore 1965*a*).

The reasons for the differences in the concentrations of residues, and their physiological significance, remained a mystery. In early 1964, Dr Moore wrote:

we do not know yet whether the unusually high concentrations of insecticide in predators are due to some common physiological characteristic or to concentration in their prey. We have no experimental evidence to show whether the amounts we find in predators are doing them harm (Moore 1964*a*).

Although proof was lacking, it seemed too much of a coincidence that the predatory species containing the highest levels of residue should be those undergoing a striking, and unprecedented, decline in reproductive capacity. As a group, predators were the most susceptible to any deleterious trend in the environment. Their population was low and their feeding habits restricted. Because they were at the end of a food chain, they might be exposed to an exceptional degree to the risks of contamination. Predators could fulfil a valuable role by acting as an early warning of imminent and harmful changes taking place.

Dr Moore wrote in December 1963 that 'the results obtained from analyses of predatory birds had done more to change scientific opinion than anything else'.[23] From the various surveys and analyses, it was clear that the most dangerous property in a pesticide was its degree of persistence. By overlooking this fact, the Notification Scheme (now called the Pesticides Safety Precautions Scheme) had failed to provide adequate protection for wildlife. Indeed, the Scheme had, unintentionally, made matters worse by encouraging the use of such compounds as DDT because of their comparatively low toxicity. No matter how carefully applied, and even if released at rates well below those likely to cause direct poisoning, persistent chemicals would soon be dispersed over a wide area. Persistence enabled poisons to build up in natural systems and have a sub-lethal effect on reproduction. Whereas most species were well adapted to deal with local increases in death-rate, they were far less able to cope with severe disruption to birth-rates.

In its evidence to the Advisory Committee in 1963, the Conservancy called for a fresh appraisal of the criteria on which the Pesticides Safety Precautions Scheme was based. Simple tests on oral toxicity could provide no guidance to the dangers that might arise from a compound's persistence. Industry should also supply information on:

(1) the cumulative effects of small doses under different physiological conditions;
(2) the sub-lethal effects on reproduction;

(3) the availability of the chemical to the animal in the field.

As well as more laboratory and field tests, there should be systematic surveys for at least a year after the new compound was introduced.[24]

By late 1963, the Conservancy was even more emphatic in its demands for interim measures to ban, or curb, the use of organochlorine insecticides. The onus should be on the manufacturers and users to prove that insecticides were harmless to wildlife. Unless they succeeded in doing this, it would be wrong to allow the environment, and particularly water supplies, to be contaminated by pesticides 'whose modes of action were obscure'.

Not only did the Conservancy begin to have original data to back up its recommendations, but industry itself showed a greater readiness to accept further curbs on the use of pesticides. In a lecture of November 1963, Dr E.F. Edson, the Director of Fison's Chesterford Park Research Station, gave two reasons why these were needed. Firstly, scientists in industry and academic spheres had failed to produce pesticides that were sufficiently selective in their impact. Secondly, the chemical industry had been so 'dynamic, fast-acting and commercially strengthened' that its pace had outstripped that of the biologist. Chemists had unwittingly created in a year biological problems of such complexity that it would take biologists a decade to evaluate and resolve them. There was strong circumstantial, and now irrefutable, evidence that persistent insecticides had inflicted considerable harm on wildlife. They posed not only 'an acute risk to birds, fish and aquatic insects, but also a more complex and longer-term threat due to their accumulation in soils, watercourses, and their selective concentration in wild species' (Edson 1964).

The partial ban

Having taken evidence from a wide variety of witnesses, the Advisory Committee completed its report in February 1964, and published it a month later. There were detailed reviews of eight compounds used in agriculture, horticulture and the home garden, food storage, industrial premises and houses, and in veterinary practice. The consequences of withdrawing any of these pesticides were given. As well as an assessment of the problems of residues in human foodstuffs, long sections were devoted to the alleged hazards presented to human health and wildlife (Advisory Committee on Poisonous Substances 1964).

The Committee rejected calls for a total ban on the use of organochlorine pesticides. There was, for example, no evidence that these compounds constituted an immediate, or serious, threat to human health. The Committee nevertheless recommended that the 'cumulative contamination of the environment by the more persistent organo-chlorine pesticides' should be arrested, and curtailed, wherever possible. Residues were found in many parts of the human body. Whilst they might be innocuous, it would be better if they could be removed altogether.

Turning to wildlife, the Committee agreed that the voluntary restrictions imposed on seed-dressings in 1961 had succeeded in greatly reducing the number of casualties among seed-eating and predatory species. The task was now to see whether further restrictions were needed to counter any long-term effects. The chief impediment to reaching a decision was lack of information on the significance of different residue levels, and the amounts of different pesticides needed to kill different species. Hardly anything was known about the effects of non-lethal doses taken, perhaps, over a long period of time. There was little information on the significance of pesticides in relation to other causes of mortality. As a result, any recommendations had to be based 'to some extent on reasoned inferences drawn in part from circumstantial evidence'. There was a need to build up an exact, comprehensive, and detailed knowledge of all the complex effects of residues, with a view to assessing the contribution made by each chemical to agriculture, its possible side-effects, and the feasibility of substituting alternative forms of pest control, where desirable. The aim should be to identify chemicals that were 'no more persistent than is necessary for effective control'. They should have the lowest possible toxicity to other species, and should not be used more widely than was absolutely necessary.

The Committee recommended a ban on the use of aldrin and dieldrin in sheep-dips, sprays, and fertilizers. Seed-dressings containing aldrin, dieldrin, and heptachlor should be used on winter-sown wheat only during the period to the end of December, and where there was a real danger of attack from wheat bulb fly. The compounds could also be used for rubbed and graded sugar-beet for precision drilling. Only commercial growers could use aldrin and dieldrin against wireworm in potatoes and to control cabbage root fly and narcissus bulb fly. All other uses of these compounds should be banned, and even these permitted uses should be reviewed after three years with a view to their being discontinued.

The Advisory Committee conceded that the recommendations were not a perfect solution. Not only might they reduce crop protection and increase labour costs, but they could lead indirectly to an increase in the use of DDT as farmers sought substitutes for banned compounds. Recent studies had shown that residues of DDT were already more widespread than had previously been thought possible. The Committee recommended that the use of DDT and gamma-BHC should be reviewed after three years, by which time it believed industry should have found even less persistent alternatives to aldrin, dieldrin, and heptachlor.

As soon as he received the report in February 1964, the Minister of Agriculture entered into discussions with the various interested parties. A member of the Ministry's Plant Pathology Laboratory later commented that the recommendations were a further, severe test of the voluntary safety arrangements. Industry was again being asked to stop supplying a large quantity of material to the British farmer. This was difficult enough to accept, but manufacturers knew that governments in many parts of the world were likely to follow the British lead. After 'tough and prolonged' negotiations, the recommendations of the Advisory Committee were accepted by industry, 'not so much because they fully agreed with them but for the sake of preserving the close relations' that existed between industry and the Government under the voluntary scheme (Miller 1965). Indeed, one manufacturer protested strenuously over the extended ban on aldrin and dieldrin, claiming that the Committee had ignored new evidence which suggested the compounds ceased to have a cumulative effect once they reached certain harmless levels of concentration. After looking at the evidence, the Committee firmly rejected this notion.

In a statement to parliament in March 1964, the Minister announced that the Government had accepted the Committee's recommendations. Assurances of co-operation had been received from all interested parties under the Pesticides Safety Precautions Scheme. A ban on most uses of aldrin, dieldrin, and heptachlor would take effect in 1965, and of aldrin and dieldrin in sheep-dips at the end of 1965. A total ban on these substances, and an extension to cover some of the less toxic substances, would be considered further in 1967.[25]

Although the restrictions were generally welcomed, there were criticisms from two directions. The National Farmers Union drew attention to the adverse effects on hill farming. In correspondence, the Minister agreed that hill farmers would be particularly affected but, in his words:

I have to strike the right balance between the practical problems which face farmers and the opposing considerations advanced in the Advisory Committee's Report. The Committee is a responsible and disinterested body, and the decisions which I have taken on the Report are not a political compromise, but the result of fully weighing the advice I have from the Committee and from the interests that stand to be affected by their recommendations.

The Minister agreed that there was little chance of a pesticide as cheap and efficient as dieldrin being discovered in the laboratories at such short notice, but, he argued, the postponement of the ban would have encouraged further delay in the development of substitutes.[26]

There were misgivings of another sort. Several members of parliament wanted to know why the restrictions had not been introduced immediately. In an Adjournment Debate, Mrs Joyce Butler spoke of great public anxiety over reports of pesticide residues being found in lamb intended for human consumption. In the conservation movement, it was widely believed that the Advisory Committee had decided to recommend a ban on dieldrin-dips because of the alarm expressed by medical experts over the level of residues found in meat being sold in butchers' shops. New Zealand had already stopped the use of such dips following a ban imposed by the United States on imports of lamb and mutton from that country because of the levels of residue arising from the use of organo-chlorine dips.

When challenged by Mrs Butler to introduce an immediate ban on such dips, a Ministry spokesman described how flockmasters would have to dip their sheep two or three times a year in order to achieve the same degree of protection as on dipping with dieldrin. They needed time to make these adjustments and, in that period, the Government hoped a new and equally effective dip might be introduced, without the dangerous side-effects of dieldrin.[27]

The Conservancy attached considerable significance to the partial ban. Dr Moore estimated that it would lead to a reduction of 70 to 90 per cent in the use of aldrin and dieldrin. Although regrettable, the short delay in banning the use of the pesticide in sheep-dips was reasonable in the circumstances. Despite very strong opposition, the Minister had taken heed of the advice of ecologists and others in reducing the scope for environmental contamination by pesticides. The withdrawals were implemented under the same voluntary scheme as had already been used to curtail the use of seed-dressings. At the Conservancy's instigation, the Ministry sought and received an assurance in mid-1965 that firms would not encourage or allow farmers to stockpile dieldrin dip in the months before the ban came into effect.[28]

WILDLIFE AND HERBICIDES

Attention was focused on those pesticides and their uses which had the most dangerous side-effects. The Toxic Chemicals and Wild Life Section decided therefore to concentrate on the organochlorine insecticides, although they accounted for less than 15 per cent of pesticides in common use. They were 'one of the star turns' because of their threat to wildlife and the wider environment (Moriarty 1975*b*). It would, however, be wrong to suggest that no thought was given to other forms of pesticide use.[1] The Conservancy expected the emphasis to change as organochlorines were replaced by other compounds. The resources previously devoted to analyses of dieldrin and DDT residues would gradually become available for studies on the effects of organophosphorus and other compounds. Whilst organophorphorus pesticides were less persistent, Moriarty (1972*b*) warned that they might nevertheless pose a significant threat to wildlife.

It was not so easy to decide whether the Toxic Chemicals and Wild Life Section should devote much attention to the use of herbicides. Dr Moore drew an important distinction between the kind of threat posed by insecticides and herbicides. Although the latter made up 80 per cent of pesticides in use, the bulk of the damage caused to wildlife was intended. Far from being a side-effect, it was the explicit purpose of spraying to remove poppies and other weeds from the cornfield. In this respect, the impact of herbicides was no different from that of the hoe and other traditional methods.

Herbicides in farming and forestry

It did not require a research programme to demonstrate that more and more plant species were likely to disappear as the use of herbicides increased. The only way to arrest such losses was to stop using herbicides.

There was no evidence to suggest that herbicides had the kind of unintended side-effects associated with organochlorine insecticides. In the early 1960s, the Toxic Chemicals and Wild Life Section carried out a study of the effects of herbicides on soil organisms. Although such compounds as MCPA had been in use for many years, it seemed prudent, in the light of experience with insecticides, to prove that there was no discernible damage to the soil fauna. Dr Brian Davis readily took up an invitation to study the immediate and longer-term effects of repeated applications of MCPA on plots of arable land on the Bridget's Experimental Husbandry Farm near Winchester. For the previous 13 years, the plots had been sprayed on 10 occasions, as part of a study to assess the effects of weed growth on cereal yields.

Davis (1965) concluded that the effects of the herbicide were so short-lived in the soil that any longer-term changes in soil fauna were likely to be expressed indirectly, namely through changes in weed flora and micro-habitat. The arthropod fauna taken from treated and control plots, both before and after spring sowing and treatment, and about a month later, indicated no differences in the size and character of the populations. No significant differences could be found in a more detailed study of the Acari population.

Much more concern was focused on the increasing use of herbicides in the plantations of the Forestry Commission and by private woodland owners. The Conservancy's regional officer for East Anglia cited several examples in 1967 of the real and potential risks to Sites of Special Scientific Interest, or adjacent nature reserves, arising from the large-scale use of 2,4,5-T (trichloro-pheno-oxyacetic acid) as a means of brush control. At a liaison meeting in January 1968, steps were taken to ensure closer liaison at the officer level between the Conservancy and Commission with respect to the management of Sites of Special Scientific Interest on property acquired by the Commission, and on land adjacent to reserves. The Commission would provide annual, and longer-term, forecasts of its spraying programmes. For its part, the Conservancy agreed to prepare a leaflet describing ways of enhancing woodland for conservation purposes.[2]

In view of mounting public concern over the effects of spraying, the Commission issued a revised memorandum to staff in 1968 setting out the procedures to be followed when using toxic chemicals. The Conservancy was consulted, and most of its suggestions were incorporated in the final draft. It was the policy of the Commission:

to use pesticides wherever they appear necessary for the most economical and efficient protection and management of our crops, *provided* that such use does not harm the crops or other interests of neighbours, destroy refuges of rare plants and animals, or affect public water supplies.

The use of herbicides in areas of high amenity had to be carefully considered so as to avoid adverse criticism. For the first time, a section of the Commission's Annual Report of 1968–69 was devoted to wildlife and conservation. Not only was the opportunity taken to announce the opening of a Wild Life Centre in Grizedale Forest, but also to explain how the Commission was taking 'the utmost care to ensure that, when using weedkillers, insecticides, and fungicides in forest management, any possible harmful effects to flora and fauna are mini-mised' (Forestry Commission 1970).

Popular concern over the side-effects of herbicides was aroused by the military use of 2,4,5-T as a defoliant in the Vietnam war, and by reports of abnormally high numbers of babies being born malformed. There was also concern over the health of American servicemen who had handled the defoliants. In October 1969, the Science Adviser at the White House announced restrictions on the use of 2,4,5-T, particularly where the herbicide might come in contact with food crops, watercourses, or highly-populated areas. The announcement was exceptional in that it was made at so high a level in Government and without preliminary discus-sions with industry (Nelson 1969). The immediate pretext for the regulations was a series of tests carried out on behalf of the National Cancer Institute, which indicated that large oral doses of 2,4,5-T, fed to rats and mice in early pregnancy, could result in deformities in their young. It was discovered later that the test chemicals, like those in Vietnam, contained high levels of an impurity, the so-called dioxin, which was known to be both highly toxic and teratogenic.[3]

Not surprisingly, these findings aroused considerable disquiet among operatives in Britain who handled the herbicide as part of their normal work. Over 80 per cent of 2,4,5-T was used in plantations of the Forestry Commission. In April 1970, the Commission announced the withdrawal of 2,4,5-T in deference to the wishes of the trade unions. In vain did both the Forestry Commission and Ministry of Agriculture stress the unim-peachable safety record of the herbicide in the years since its introduction to Britain in 1956. They emphasized that the American restrictions merely brought that country into line with British practice. In America, it was still possible to use 2,4,5-T in controlling weeds and brush in forests and other non-agricultural land. The levels of dioxin in the herbicide used in Britain were as low as 1 ppm, compared with levels of 27 ppm recorded in the American trials.

In view of the number of compounds which the Conservancy wanted to see withdrawn, it was perhaps ironical that so much public concern was focused on this particular herbicide. It was one of the few chemicals that was used by conservationists in the management of their nature reserves. The Toxic Chemicals and Wild Life Section had studied the effects of 2,4-D and 2,4,5-T on the High Halstow National Nature Reserve in Kent, and at Wicken Fen. When applied to cut stumps, there did not appear to be any adverse effects on either the plants or soil fauna. The cleared areas were soon recolonized, often with species absent from the uncleared parts of the reserves. Far from harming communities, the results were sufficiently encouraging for an experiment to be designed for controlling the re-growth of oak, chestnut, and hornbeam in a newly established wayleave for power cables on the Ham Street Woods National Nature Reserve. It was hoped to arrest growth at a predetermined height, so as to provide a scrub habitat for birds and insects on a reserve where this kind of habitat was poorly represented.

In view of the lead given by the Forestry Commission, an urgent reassessment of the safety of the herbicide was made by Dr Way of the Toxic Chemicals and Wildlife Section in late 1969. He was able to assure regional staff that it was highly improbable that ground applications of 2,4,5-T at recommended rates of application for woody plant control could produce a sufficiently high concentration of residue in foilage, soil, or water to constitute a toxic hazard. Except for deciding what should be the maximum permissable level of dioxin in 2,4,5-T, the Advisory Committee on Pesticides concluded in May 1970 that no change was required in the existing use of the herbicide.[4]

The Nature Conservancy was alarmed, not so much by the threats posed by individual herbicides, but by the increasing scale of herbicide spraying, both on intensively used and marginal farmland. For this reason, it welcomed the decision of the (Frazer) Research Committee to issue a supplementary report, dealing specifically with research needs in the herbicide field (Agriculture Research Council 1965). In that report, the Committee spoke of its concern that many herbicides had been introduced with little knowledge of their effects on the environment. The dangers inherent in this practice were underlined in an article in the *New Scientist* by the Director of the Weed Research Organisation of the Agricultural Research Council. Under the title 'The herbicide revolution', Fryer (1964) described how, over the course of 20 years, weedkillers had developed from garden aids and agricultural curiosities into an important branch of the chemical industry and a major factor in crop production.

An important turning point had been reached in the development of herbicides. The emphasis so far had been on incorporating them into a system of husbandry that had remained basically unchanged since the time of Jethro Tull and the introduction of the concept of sowing crops in rows as an aid to horse-hoeing in the eighteenth century. Herbicides had merely taken the place of the hoe, harrow, and sickle. There was now evidence of herbicides being part of a much more fundamental change in crop husbandry. Because of their unique ability to kill vegetation on a large scale without disturbing the soil, ways were being sought to use herbicides to change the composition, and thereby improve the productivity, of pastures and meadows without recourse to ploughing. Herbicides were being used to keep the soil beneath orchard and soft-fruit trees clear of vegetation, and there was the increasing prospect of being able to grow a succession of cereal crops without recourse to the plough or any other form of soil cultivation. Through the development of a special drill, seed could be sown directly into any stubble and weed growth already killed by spraying. The sparser and less varied weed flora would be increasingly confined to other parts of the countryside.

Not only might some weed species disappear, but the (Frazer) Research Committee warned that those animal species dependent on weeds for food or shelter would also become scarce. The trend might have started already in East Anglia, where the local elimination of weed species had led to a decline in the insect population, which had in turn deprived partridge (*Perdix perdix*) chicks of essential proteins at a critical stage in their development. Clearly, the effects of large-scale herbicide use on wildlife, in the widest sense, were unknown. Dr Fryer suggested that it was the responsibility of the herbicide specialist, acting as a member of a team, to assist in assessing any harmful effects on wildlife and the soil. Clearly, there was need for an ecologist in the team appraisal.

Dr Moore continued to insist that 'a major onslaught on the ecological effects of herbicides on ecosystems would be a mammoth undertaking and would not, in our present view, produce returns commensurate with the resources which it would require'.[5] An extensive search through the world literature for reports of toxic hazards from the commonly used auxin herbicides (including MCPA and 2,4-D) had failed to produce any authenticated instances of direct or indirect toxic hazards to wildlife from the correct application of chemicals in agriculture (Way 1969*b*). The hazards posed by urea, diazine, and triazine herbicides were no greater, and might even be less, than those from an application of a similar weight of common salt to the ground (Way 1968).

Even if the resources for such an extensive study of the ecological effects of herbicides had been available, without jeopardizing the effort being made in other fields of pesticide research, it was doubtful whether any demonstration of harmful effects of herbicides on wildlife would have led to any significant reduction in usage. Many farmers saw them as the only method of achieving greater efficiency and food output. There was unlikely to be any decline in use until farmers began to look more closely at the cost-effectiveness of using herbicides (Way 1965). It was the agronomist rather than the ecologist who was most likely to bring about this change in attitude.

There were some signs of this reappraisal taking place. By the mid-1960s, over 80 per cent of all crops were sprayed; MCPA was so cheap to apply that it was used prophylactically on many fields as a matter of routine, primarily to obtain a clean crop at harvest time rather than for any increase in yield. Because the fields had often been sprayed many times already, the weed seed population was already far too low to compete significantly with the crop. At the Weed Control Conference of 1964, the former Director of the Weed Research Organization suggested that the time might have come for more research to be devoted 'to find-ing out *when* it is economic to spray' (Woodford 1964).

These discussions were symptomatic of large-scale changes taking place in the management of farmland, including the adoption of new rot-ations, greater use of clean seed, combine harvesting, and the destruction of hedges and 'waste' land. Whether taken singly or together, these trends provided the conservationist with further evidence that game and wildlife could no longer be regarded as the automatic by-products of farming. Instead of trying to curb the use of herbicides, conservationists had to make a fundamental reappraisal of the countryside and strive towards a new pattern of land use and management, where some areas were zoned primarily for agriculture and others for conservation purposes (Moore 1970c).

Herbicides and aquatic systems

The significance of herbicide use on agricultural land was enhanced by what was happening to the vegetation around bodies of water and along-side railways and roads. No longer could the conservationist assume that the river bank, railway cutting, or roadside verge would provide refuge for species displaced from farmland. There was growing evidence of herbicides being used on an increasing scale for non-agricultural pur-poses, and therefore outside the aegis of the Pesticides Safety Precautions Scheme.

There were soon pointers to the problems that lay ahead. In June 1961, the regional officer of the Nature Conservancy in the East of England discovered plans to spray sections of the Norfolk Broads by helicopter. In Dr Duffey's words, the proposal constituted one of the gravest threats to conservation in Norfolk for many years. The spraying was intended merely to help anglers move along the banks, and it had not occurred to the River Board to seek the advice of, let alone forewarn, the Con-servancy. In the event, the threat came to nothing, largely because no insurance company would indemnify the River Board. The Fisheries and Pollution Committee of the River Board decided that the benefits of spraying would hardly justify the risks to boats and holidaymakers.[6]

Meanwhile, there was mounting concern in another famous wetland area, the Romney Marsh, where the River Board was also experiencing increasing difficulties in controlling the growth of vegetation. As early as 1952, the engineer to the Board had described how the labour situation made it 'well nigh impossible to recruit new men who are capable of undertaking weedcutting operations in a wholly efficient manner'. It was impossible to cut and clear all the weed growth in the short period avail-able at the end of summer. The first trials with a selective weed-killer,

Dalapon, were made by the suppliers in May 1959 and by 1961 the engi-neer described spraying as a useful adjunct to the traditional forms of weed clearance. A timber-framed aluminium boat was specially adapted for spraying purposes and a landrover and knapsack spray were used for the more inaccessible places. In 1967, a helicopter was used, under con-tract from the Shell Chemical Company, to spray emergent weed with Dalapon and 2,4-D.[7]

When the Kent Naturalists' Trust protested about the threat to wildlife in Romney Marsh, it was reminded of the River Board's statutory responsibility for keeping watercourses clear of weed. For practical pur-poses, the engineer assured the Trust that only a minute proportion of the watercourses would be sprayed, and that there would remain 'plenty of weed-choked channels to delight the botanist'. When Dr Moore visited the area in September 1962, he found spraying taking place on a consid-erable scale. The Trust had 'every reason to be concerned', but there was a risk of exaggerating the impact of spraying. As Dr Moore pointed out, Dalapon was one of the least toxic materials used and the Conservancy could not possibly press for its immediate suspension in view of the real difficulties faced by river boards.[8]

Three forms of vegetation control were practised, he continued, namely, against emergent plants (mainly reed, reed-mace and bur-reed), submerged plants (like starwort and pond weed), and algae. It often hap-pended that the removal of the dominant monocotyledons led to bad infestations of submerged weed and algae. Because of strong angling interests, it was very unlikely that river boards would attempt to elimin-ate all life in the watercourse. In an appraisal of the problem, Dr Moore warned that the chief danger was to slow-moving, unfished waters in the lowlands, although it was probably wise to asume that spraying would always lead to changes in plant and animal life, wherever applied. Even if chemicals were relatively non-toxic, the presence of large masses of rott-ing vegetation would lead to de-oxygenation of the water. In the longer term, striking changes in the habitat might occur. The demise of a large reed bed could lead, for example, to the displacement of the reed war-bler. The most critical need was to prevent spraying operations becoming so extensive as to exterminate restricted populations of wildlife along the watercourses.

It was the lack of any kind of administrative procedure that most wor-ried the Nature Conservancy. In correspondence with the Conser-vancy in 1961–62, the Ministry of Agriculture drew attention to the way in which the National Association of Agricultural Contractors had emphasized, in its Code of Conduct, the need to avoid taking risks when carrying out spraying operations. The Ministry had asked for these dangers to be stressed even more strongly in a revised edition of the

Code in 1962. These assurances could not, however, provide a guarantee of protection. Even where some form of rudimentary liaison was established between those responsible for the watercourses and a conservation body, set-backs could occur. In 1963, a press bulletin of the Council for Nature described how a length of canal, set aside by the Leicestershire Naturalists' Trust as a nature reserve for schoolchildren, had been sprayed in error. Spraying should have been limited to that needed to keep the canal open for navigation. The Conservancy's files contain many other perceived threats, and recorded disasters, arising from herbicide applications in the freshwater habitat.

When he came to draft a set of regulations, Dr Moore stressed the urgent need to draw up agreements with all the relevant interests at a national level. Both the Conservancy and Naturalists' Trusts should acquire more lengths of watercourse as nature reserves, and greater protection should be accorded to existing and potential Sites of Special Scientific Interest. Procedures should be drawn up whereby the river boards consulted the Conservancy before any sprays were used within, or near, such areas. Where herbicides were used, a code of spraying should be followed. An entire pond or length of river should never be sprayed at one time and only the smallest quantities of spray should ever be used.

In a letter to the Office of the Minister for Science in 1962, the Conservancy stressed the need for a new initiative. It was extremely doubtful whether the activities of the river boards, and other bodies operating in a non-agricultural context, were subject to the regulations of the Pesticides Safety Precautions Scheme. Even if they did fall within the terms of reference of the Advisory Committee on Pesticides, the latter had no resources to investigate them. As Dr Moore had commented in November 1962, the issues arising from the use of aquatic herbicides were 'quite different from the normal notification one since useful advice could only come from knowledge of local conditions; knowledge of the effects of the chemicals was only of subsidiary importance'.[9]

Contact was made in 1963 with the Fisheries Department of the Ministry of Agriculture and with other relevant research organizations. Soon the question of regulating the use of herbicides on watercourses merged with anxieties over the effects of insecticides used on neighbouring farmland. In its report of 1964, the (Frazer) Research Committee called for even closer collaboration between the Conservancy and other interested parties in assessing the contamination of freshwater by pesticides, the accumulation of residues in animal life, and the way in which these might bring about changes in the fauna and flora of acquatic systems.

An opportunity for the Toxic Chemicals and Wild Life Section to study the effects of herbicides on freshwater systems arose in 1964, following an invitation from Plant Protection Ltd to participate in trials on the effects of diquat and paraquat on submerged water weeds. Considerable experience was gained in recording aquatic vegetation and invertebrate populations. It was found that although submerged weed was severely affected by herbicides at concentrations of 0.5 to 2.0 ppm, more than one application was usually needed to prevent re-growth. There was no significant effect on algae. Vegetation growing on the bank at the waters' edge was generally unaffected except where it actually grew in the water. Perhaps the most significant lesson to be learned from these trials by the Section was the extent to which herbicides became diffused through the water—an important consideration where stretches had to be kept free from herbicides in order to protect plant and animal communities of high biological interest.

In the light of this experience, a collaborative exercise was mounted on two of the five lakes at Oxton in Nottinghamshire in 1965–66 by the Toxic Chemicals and Wild Life Section, the Jealotts Hill Research Station of ICI (which periodically analysed the mud, water, and vegetation for residues of paraquat), and the Water Pollution Research Laboratory at Stevenage (which took a number of physiochemical measurements and placed captive fish in the lakes for observation). In 1963, the relatively clear water in the shallow gravel and sand pits had become covered by a thick growth of Canadian pondweed (*Elodea canadensis*). This, together with large mats of algal scum, had made fishing almost impossible, and the local fishing club had applied to the Weed Research Organisation and Trent River Board for advice on how the vegetation cover might be removed.[10]

A jointly written paper (Way, Newman, Moore, and Knaggs 1971) described how trials with paraquat eliminated most of the primary producers, both phanerogams and algae, although the effects on different groups of algae varied considerably. Except for *Polygonum amphibium* and *Chara* spp., all the submerged and floating plants were eradicated within 32 days of application, and one of the lakes treated remained substantially free of vegetation for two years. Although the principal submerged weed had been *Elodea*, the most likely recolonizers were found to be *Potamogeton crispus* and *P. pectinatus*. The breakdown of such large quantities of weed *in situ* led to severe de-oxygenation, and captive trout in cages died. Free-living coarse fish and trout were severely stressed for a short period in the first year of the trial, but recovered and were generally unaffected in the second year.

Despite the profound changes brought about by the loss of most plant life, neither aquatic invertebrates, fish, nor breeding birds appeared to suffer obvious changes in species or population during the period of observation. The herbicide was rapidly absorbed from the water by the vegetation, significant quantities of the compound being present in plants within 24 hours. A gradual build-up of paraquat residues in bottom deposits was recorded at 32 days, and subsequently 197 days, after application, with a marked fall-off at 364 days.

Meanwhile, the Fisheries Department of the Ministry of Agriculture had set up a Research Group to review and collate research on 'Pesticides and the Aquatic Environment'. At a meeting in February 1964, the various interests concerned with the management of watercourses had agreed on the need for 'planned practical experimentation', where the effects of the use of agricultural chemicals on standing and running water could be measured. The Research Group was intended to study the impact of the longer-term changes, and to identify the various ways in which residues entered the ecosystem.[11]

Despite these various moves, the regulation of aquatic herbicides continued to be 'one of the most urgent conservation problems'. By the early 1970s the river authorities were moving from a position where herbi- cides were used only as a local basis to one where extensive and adjacent areas were being treated. Very few authorities appeared at that time to recognize the biological significance of the trend, and that it might become impossible for plant and animal species to recolonize treated waters. Yet in practice, the only people capable of providing engineers with detailed advice on scale and methods of use were the biologists employed by the authorities. For this to happen, there had to be closer collaboration between the two parties, and a considerable improvement in the resources accorded to the biologists for studying such large environmental issues. These were questions that clearly transcended the herbicides issue.[12]

Railway banks and cuttings

Concern for the future of wildlife on railway banks and cuttings was stimulated in April 1961 by a Parliamentary Question tabled by John Farr, which asked the Minister of Transport to stop the British Transport Commission from carrying out a scheme to use toxic sprays on 3000 miles of railway line. The Minister refused to intervene on the grounds that 'British Railways use only non-toxic weed killer'. Dr Moore nevertheless urged the Conservancy to seek further information about the scheme as 'a matter of great urgency and importance'. In view of the season of the year, it was 'quite conceivable that even a few days delay might be disastrous'.[13]

In a memorandum, Dr Moore described how the grass banks and cuttings owned by British Railways included a wide range of habitats. Many were rich in species, and had remained relatively undisturbed since the time of their construction. A recent survey by the Toxic Chemicals and Wild Life Section suggested that railway verges might often be richer in species than roadside verges. Using the cowslip (*Primula veris*) as an indicator species, 10 colonies (360 plants) had been found along a length of 25 miles of roadside in Huntingdonshire whereas 6 colonies (1000 plants) had been found on seven randomly selected lengths of railway, each one-third of a mile long. At a time when modern agricultural techniques were leading to the destruction of permanent pastures and hedgerows, the conservation value of the banks and cuttings had greatly increased. In many districts of eastern and midland England, they provided some of the last resorts for some plant and animal species. The survival of these reservoirs for wildlife could not be taken for granted; everything depended on existing management policies being continued by British Railways.

It was likely that the Parliamentary Question was referring to the use of special trains for spraying a width of up to 7 feet from the track with a weed-killer. The trains were first used in 1960, and a member of the Toxic Chemicals and Wild Life Section, who had made a short journey on one of them, reported that every care was taken to prevent spray from drifting. At present, there was no damage to the flora and fauna of the banks and cuttings, but Dr Davis warned that 'there is a very real danger that spraying will be extended from the immediate area of the rails to that of the embankments and cuttings in the future'.

In July 1961, the chairman of the Nature Conservancy, Lord Hurcomb, wrote to the British Transport Commission (of which he was also a member), asking for confirmation that no change in spraying practice was contemplated and that the Nature Conservancy would be consulted if, at any time, it was intended to spray a wider area. Whilst the Conservancy appreciated the great difficulties faced by British Railways in managing the railway formations, the banks and cuttings were nevertheless 'a national asset, both from the scientific and other points of view' in providing a 'protected habourage' for wildlife. The Chairman of the British Transport Commission replied by assuring the Conservancy that only the track and cesses would be sprayed. The amount of disturbance to the 'natural growth on the railway slopes would be minimal: non-toxic weed killers would be used on 'a very limited scale' where grass growth might interfere with cables. In one or two instances, a brushwood killer might be used to eliminate brambles as part of rabbit clearance schemes.[14]

There the matter rested. During the remainder of the 1960s, attention was largely focused on finding ways of using redundant railway lines for nature conservation purposes. In a paper on the disposal of abandoned railway lines in 1963, the British Rail Board expressed its readiness to set up a regular procedure whereby the Nature Conservancy would be informed of the lines for disposal in the hope that negotiations might be opened 'to the mutual advantage of both bodies'. Thereafter, lists of closed lines were sent to the Conservancy, which passed them on to the voluntary bodies.[15]

In the triennial Monks Wood report for 1969–71, Dr Mellanby again drew attention to the ecological importance of the embankments of operational lines. There were protests from local naturalists when an extensive area of grass and scrub between Cheltenham and Honeybourne was burnt at nesting time. This episode encouraged demands that the licence which permitted British Rail to burn their slopes during the nesting season should be revoked.

In a letter to the Nature Conservancy, the Chairman of British Rail described the changing needs of vegetation management on railway property. Until the late 1960s, a swathe of grass was traditionally cut at the top and bottom of the slopes, and the grassland covering the remainder was burned two or three times a year. This helped to stop grass from seeding, and the seeds from drifting on to the cess and track ballast where, in time, weeds would grow and impede drainage. Controlled burning and cutting also helped to reduce the chance of sparks from steam trains causing accidental fires, which could easily spread to neighbouring properties. To facilitate this very necessary management work, the Ministry of Agriculture had issued a licence (Statutory Instrument 1949, 495) which allowed the slopes to burned at any time of the year, thereby excluding British Rail from the Provisions of the Heather and Grass Burning (England and Wales) Regulations (Statutory Instrument 1949, 386). During the 1960s, the introduction of spray trains had virtually eliminated any risk of weed growth in the cess and track ballast, and the replacement of steam engines had greatly diminished the risk of accidental fires, but, in his letter to the Conservancy, the Chairman of British Rail insisted that the licence was still required for controlled burning throughout the year. Too much scrub and grass growth would threaten the safe-working of the railway, and provide cover for such pests as the rabbit and fox

In the light of this information, the Conservancy supported British Rail in opposing any move to rescind the licence for year-round controlled burning. Not only was it necessary to clear scrub in some parts for safety reasons, but controlled burning would help to conserve those grasslands which had previously been maintained by fires caused by the sparks of steam-engines and by the grazing effects of rabbits before myxomatosis. In a memorandum of 1975, the Conservancy concluded that burning was an acceptable form of management from the wildlife point of view so long as 'very large stretches are not treated at the same time'. It was important to retain a mosaic effect in the vegetation.[16]

The formulation of management policies was nevertheless still entirely at the discretion of British Rail, and the Conservancy continued to press for closer liaison over the timing and location of work. Even greater fears were aroused following reports in 1975 of British Rail awarding a three-year contract for spraying scrub with herbicides. The *Daily Telegraph* published a report of paraquat being used on a length of railway embankment at Berwick-upon-Tweed. British Rail responded by describing the spraying as an experiment, adopted because of the impossibility of finding sufficient staff to carry out the more traditional methods of control. A month later, in October 1975, the Chief Civil Engineer of British Rail told readers of the *Guardian* that the control of vegetation was 'now mainly achieved by using weedkillers and growth inhibitors'.

In an internal review within the Conservancy, everyone agreed that it would be a tragedy for nature conservation and aesthetic reasons if a new policy for the widespread use of herbicides and/or growth retardents was introduced. On the other hand, 'a small amount of selective spraying would not be harmful from the national point of view *if it could be controlled*'. In a further letter to British Rail, the Conservancy again proposed closer liaison, adding that 'since it is clear that some railway verges are more important than others as reservoirs of wildlife . . . it may be practicable to reach an agreed schedule indicating the kinds of and types of verges which should not be sprayed and conversely those which could be sprayed without detriment to nature conservation. In these latter cases, it would be valuable to have agreement about the kinds and frequency of spraying'.[17]

Throughout these discussions, both parties had been conscious of the lack of basic ecological information about the characteristics of railway banks and cuttings. The Toxic Chemicals and Wild Life Section had pressed for the opportunity to obtain this information, but it was not until 1976 that an initiative could be taken. Following the decision of the Civil Engineers' Committee of British Rail to assist in selecting a number of suitable sites for detailed field investigation, the Chief Civil Engineer expressed the hope that the studies would lead to a low-cost treatment of railway embankments being recommended that would 'satisfy both the railway and ecological requirements'.

Motorway and roadside verges

Interest in the effects of herbicides on the wildlife of roadside verges was further stimulated in the early 1960s by the large-scale introduction of a growth retardant, maleic hydrazide (MH), and the technique of combining this with a selective weed killer, either 2,4-D or MCPA. Far from this being an occasional form of treatment, there was every prospect of routine applications being made to complement, if not, replace roadside mowing. At the same time, the construction of a completely new system of motorways from 1959 onwards created not only considerable disturbance over long stretches of countryside but an outstanding opportunity for what was termed 'creative conservation'.

In consequence of these trends, a four-fold programme evolved in the Toxic Chemicals and Wild Life Section, consisting of:

(1) a survey of the extent and character of roadside verges;
(2) management experiments;
(3) greater liaison with the highway authorities;
(4) a separate motorway study.

In order to assess the biological importance of the estimated 500 000 acres of roadside verges in the United Kingdom, a sample survey was made in 1967 which indicated that over 400 different species occurred on the verges, and a further 240 species at the boundary with the hedge or ditch. This represented a quarter of the British flora, and subsequent surveys raised the proportion still further. In addition, the survey identified 20 of the 50 mammals indigenous to Britain, all of the 6 British reptiles, and 40 of the 200 species of birds (Way 1970a).

If management proposals could be devised that proved acceptable to the relatively small number of authorities involved, and that helped to sustain and develop the biological interest of the verges, a major contribution would have been made. Of the £135.7 million spent on road maintenance in the United Kingdom in 1967, about 1.5 per cent was spent on grass cutting. This represented twice the Conservancy's annual budget. As Dr Way stressed, there was no disagreement on the need for management. Indeed, the conservation of tracts of rich, floriferous grassland habitat from scrub invasion was one of the best reasons for management. The differences arose over the way management should be conducted.

For the highway authorities, the main objectives of vegetation management were to maintain the stability of the roadside formations, ensure the safety of road users, eliminate weeds, and enhance amenity. The latter was usually interpreted as keeping the verges tidy in a suburban sense. To achieve this, great stress was laid on keeping the vegetation short, at least within the immediate vicinity of the road itself. From 1965 onwards, Dr Way set up trials on the chalk soils at Ickleton, near Cambridge, and on heavy Oxford clay soils at Keyston, to the west of Huntingdon. On the basis of 20 chemical and mowing treatments applied, the aim was to identify economic programmes which would satisfy the engineers and be sympathetic towards nature conservation and amenity.

The trials, together with a further experiment set up in 1968 on the M1 in Leicestershire, demonstrated the significance of understanding the growth pattern of grasses. Under average conditions, there was a rapid increase in height in May and June, reaching a peak in June and July. If the flowering stems were cut in May and June, there was no re-growth and the height of the vegetation thereafter reflected the development of the basal leaves, which did not normally exceed 12–14 inches (30 cm) in height. Comparatively little advantage was obtained from more frequent cutting (Way 1974).

There was little point in carrying out surveys and trials if the results were not relayed to those directly responsible for the management of the verges. A symposium organized by the British Crop Protection Council and Nature Conservancy in March 1969 provided an opportunity not only to exchange information and experience but, above all, to improve liaison between the various interested parties. The papers and proceedings were published (Way 1969a) and, in 1973, Dr Way compiled a further report collating and summarizing the information on objectives and methods of management, as indicated by the personnel of each of the County Council highway departments during visits made in 1972 (Way 1973). A year later, a further report was published this time outlining the extent of collaboration between the highway departments and conservation bodies in the management of rural road verges and Sites of Special Scientific Interest (Way 1974).

It was hoped that eventually highway authorities would receive lists of all the sites of biological interest in their respective areas, together with management prescriptions. The road maps of each divisional area would be marked in four different ways to identify:

(1) sites of special biological interest, perhaps requiring individual management;
(2) lengths of roadside requiring seasonal management to sustain or encourage herb-rich grassland communities;
(3) lengths of roadside where scrub might be allowed to develop;
(4) lengths of roadside with no known biological interest

Interest in the conservation of roadside verges had meanwhile been stimulated by the motorway-building programme. By 1970, almost 600 miles of motorway had been opened, 350 miles were under construction, and a further 400 miles proposed. Each mile of motorway absorbed about 200 acres of land, of which at least 40 per cent became verges, banks, and cuttings covered initially with a standard mixture of cultivated grasses and clover. The potential contribution of these areas to nature conservation was first put forward independently by Dr Moore at a conference on roads in the landscape, convened by the Ministry of Transport and the British Roads Federation in July 1967, and by Clough Williams-Ellis, a member of the Landscape Advisory Committee of the Ministry of Transport. In a booklet on *Roads in the Landscape,* Williams-Ellis (1967) commented:

with the introduction of more intensive and cleaner systems of farming much of the British flora is being ousted from meadow, pasture and hedgerow, and it seems logical that the roadside verge, particularly on the motorway where pedestrians are forbidden, should become a nature reserve and provide areas and secure habitat for our wild flowers.

In her opening address to the 'Roads in the Landscape' conference, the Minister of Transport, Mrs Barbara Castle, took up the point by Dr Moore that the landscaping of motorways provided the opportunity to win back some of the wildlife habitat that had been lost. By establishing habitats which were self-maintaining, 'the need for artificial maintenance along our roads' could be reduced (Moore 1967c). In an article in *New Scientist*, Dr Way drew attention to the fact that the estimated 10 800 acres of motorway verge covered a wide range of topographical situations and were the ultimate responsibility of a single body, the Ministry of Transport. Whilst they would not become outstanding for their biological interest overnight, it was hoped that they would come to support viable populations of grass and scrub species, representative of the countryside through which they passed (Way 1970b).

The key to their enhancement as reservoirs for wildlife, he continued, was the constancy of their management. The aim should be to create stable, yet diverse, conditions in which a varied range of plant and animal populations would be established. The methods chosen had to be accommodated easily into the maintenance programmes laid down by the Ministry of Transport. A report was published in 1976, collating information supplied by the Department of the Environment (which had absorbed the Ministry's responsibilities by that time) and the County Councils on the variety, objectives, and methods of management of grassed and planted areas beside motorways (Way 1976).

A considerable fund of knowledge was thus acquired on the biological interest of verges and the impact of potential and actual programmes of management. The very act of gathering and disseminating the information, and the closer liaison that arose out of the exercise, may themselves have had a significant influence on management policies. By 1974, almost all the highway authorities had ceased to use chemical sprays on a large scale because of the opposition of conservationists, the cost of the chemicals and their application, and doubts as to the effectiveness of the treatments. It became increasingly clear that the effects of an application of growth retardant, even using a selective weed-killer, did not last all season. Two applications were usually necessary, with perhaps a tidying-up cutting-operation at the end of the season.

A less encouraging trend for the conservationist was evidenced by the decision of an increasing number of authorities to cease all forms of vegetation management on economy grounds. Instead of being managed too 'intensively' from the conservationists's point of view, there was a danger of many herb-rich areas becoming swamped by the unchecked growth of bramble and scrub. The pendulum began to swing too far in the opposite direction (Way 1977).

THE FURTHER REVIEW OF
ORGANOCHLORINE INSECTICIDES

Pesticides pay no heed to political frontiers: both scientific and administrative aspects had to be considered on a world-wide basis. Britsh
scientists had much to learn from the trials and observations of colleagues in the United States. One of the most influential studies of the
hazards likely to arise from chemical pest control was written by Robert
L. Rudd for the US Conservation Foundation. This book was completed
in September 1963 and was called *Pesticides and the living landscape*
(Rudd 1964). 'In its quieter way', it was regarded by many as being every
bit as devastating as *Silent spring*. A foreword to the English edition was
written by Norman Moore (Moore 1965c).

The European dimension was also important. British ecologists
believed there was mounting evidence of summer migrants being affected
by the use of pesticides in their wintering grounds, and in areas through
which they passed en route for Britain (Moore 1965d and 1970d).
Recognizing the international dimensions of the pesticide problem, the
International Union for the Conservation of Nature (IUCN) set up a
commission in 1961 to study the ecological effects of chemical controls.
In 1962, a Committee of Experts for the Conservation of Nature and
Landscape was appointed by the Consultative Assembly of the Council
of Europe. This Committee appointed three working parties, one of
which covered *inter alia* the use of toxic chemicals.[1]

The personnel of the Nature Conservancy were uniquely qualified to
act on behalf of these international bodies. The Toxic Chemicals and
Wild Life Section was the only unit in Europe devoted entirely to long
term studies of the effects of pesticides on natural ecosystems. As secretary of the IUCN commission, and at the instigation of the Council for
Europe working party, Dr Moore drew up an agenda for action. As rap-
porteur of the working party, Mr R.E. Boote made considerable use of
the agenda in drafting a memorandum in June 1963 for submission to the
Committee of Experts.

In correspondence, Boote admitted that the memorandum was largely
based on British experience but, he continued, 'our experience and lines
of approach are in advance of those of many of the other 16 countries in
the Council of Europe'. The memorandum advocated the setting up of an
information centre and the adoption of schemes analogous to the British
Agricultural Chemicals Approval Scheme and the Pesticides Safety Pre-

cautions Scheme. Copies of these schemes were attached to the memorandum.[2]

At its second meeting in November 1963, the Expert Committee adopted the memorandum, and called the attention of the Council of Ministers to the need for adequate control over the selection and use of pesticides in member countries. The Committee recommended that each country should co-operate with the IUCN commission in collecting information on regulations in force and on the types of research under-way, as a first step towards establishing a European Convention on the subject.

The NATO Advanced Study Institute

It was collaboration on an international scale that provided the occasion for two important meetings in 1965. The first was an international gathering of scientists held at Monks Wood, the proceedings of which were published as a special supplement to the *Journal of Applied Ecology*. In an appraisal of the discussions at the meeting, Moore (1966*a*) wrote that participants gained a much better perspective on the problems and had become convinced 'of the paramount importance of the role of persistent organochlorine insecticides in the environment. The second gathering was convened by Professor Joseph Hickey at the University of Wisconsin in September 1965. It focused more specifically on the plight of the peregrine falcon.

The way in which Hickey's meeting came to be convened underlined the international dimensions of the pesticide issue. In a letter to the Director-General of the Nature Conservancy in August 1963, he recalled that it was five years since he had first encountered 'a silent spring' at Kenilworth, Illinois, and had begun research on the effects of insecticides on birdlife in the Middle West. He found that, although the very heavy use of DDT had led initially to mortality rates as high as 90 per cent among urban birdlife, alternative methods of control had been found and applied. The residential communities of Wisconsin had found it possible to keep their songbirds and, at the same time, to control the bark-beetle vectors of Dutch elm disease.[3]

Now, after five years, Hickey felt 'a growing sense of alarm'. The British reports of a marked decline in raptorial species seemed to be running parallel with results from small-scale studies of the bald eagle and osprey in the United States. For the first time, he felt apprehensive that 'some large-scale population changes in certain avian species may be actually taking place'. Evidence of DDT residues being found in the northern wilderness and in pelagic fish in all the oceans was extremely disquieting. Judgments that, say, 5 lb of DDT per acre killed, but 1 lb did not, seemed no longer quite so 'safe'. Meanwhile, one result of curbs being imposed on the use of persistent toxic chemicals in the United States appeared to be an increase in the export of these chemicals. Hickey wondered what would happen in 'undeveloped' countries when they embraced the 'Chemical Revolution' and allowed the compounds to enter their 'vulnerable habitats'. In his letter to Mr Nicholson, he emphasized the need for a monitoring system on a world-wide basis, with adequate analytical facilities.

Having read Derek Ratcliffe's paper in *Bird Study*, Professor Hickey organized a 14 000-mile journey through the United States in the spring of 1964, in order to check the peregrine eyries visited in a survey of 1939–40. Not a single fledgling was found between Georgia and Nova Scotia. The news was 'so dramatic as to make an international conference on the population biology of *Falco peregrinus* an absolute necessity' (Dunlap 1981). An impressive array of European and American experts, representing all facets of avian biology, sought to crystallize and evaluate all the hypotheses that might account for the decline (Hickey 1969; Lowery 1969). Ratcliffe (1980) described the meeting as by 'far the most exciting scientific meeting I have ever attended, with a clear goal, and I believe it had a tremendous effect in stimulating various lines of work on the conservation of the Peregrine'.

The meeting at Monks Wood in July 1965 was the first to bring together scientists from many parts of the world who were working on 'different but complementary aspects of pesticide research'. It had resulted from moves made since 1963 by Dr Moore, who had approached NATO in his capacity as secretary of the IUCN committee. NATO agreed to sponsor an Advanced Study Institute on 'Pesticides in the Environment and the Effects on Wildlife', with Dr Moore as director. Attendance was restricted mainly to government and university scientists, who attended in a personal capacity. They came from eleven countries, both inside and outside NATO. Because the Institute was intended to have an educational role, a balance was struck between those with considerable experience, and those who were just starting to work on pesticides (Moore 1966a).

The Institute was attended by 71 scientists, and lasted for the comparatively long period of 12 working days, interspersed with excursions to other research centres and a national nature reserve. Thirty four papers were read, many of which took the form of progress reports, and, between them, they reviewed most aspects of the pesticide problem in the terrestrial, freshwater, and marine environments. In order to allow plenty of time for discussion, no more than four papers were given in one day, and much of the time was spent in informal discussion groups. Meals were provided on the Station and a bar licence was secured! Visitors from industry, universities and government organizations were invited to an Open Day, and a number of participants formed a panel and answered visitors' questions. The gathering was to mark the beginning of a tradition of informal symposia at Monks Wood, covering a wide range of topics.

In a general statement at the end of the meeting, participants noted that pesticide residues had been detected in a wide range of physical and biological samples taken from diverse environments. Their effects on wildlife were usually unknown. Participants believed that there was a particular need for the routine collection of data on pesticides and changes in the environment, experimental research on a wide range of pesticides in different environments, a greater application of existing knowledge, and a wider dissemination of that knowledge.

Despite the great volume of work in progress, everyone was impressed by how little was known in all branches of the subject. There was a continued need for basic information about pesticides in the environment and about wildlife populations. For commercial, administrative, and political, rather than scientific, reasons there were no accurate statistics on pesticide use in most countries. Moreover, little was known about pesticide abuse, or about what happened to the chemicals after application. In general, little was known also about the toxicological significance of residues and only qualified answers could be given to such questions as 'what levels of pesticide in an organ were indicative of poisoning?', and 'which organ should be analysed in comparative studies and in making assessments?'. In searching for answers, considerable attention was being focused on resistance, environmental contamination, and biological controls. There was renewed interest in the ecological aspects of population dynamics (Moore 1966*a*).

This Chapter will focus on the part played by the Toxic Chemicals and Wild Life Section in promoting research on the effects of pesticides on wildlife, and on how the results of that research were perceived by critics and by the Advisory Committee on Pesticides in a further review of organochlorines between 1967 and 1969.

The choice of indicator species

Many of the changes taking place in the environment were so subtle that they could only be discovered, and their significance assessed, if scientists set out explicitly to find them. Even then, the ecologist might still be caught unawares. The study of the peregrine falcon had highlighted the number and range of relationships that might exist between a species, other species, and its habitat. Some changes might be so complex and unobtrusive in character that they might be perceived only when they

were well underway. In practice, the ecologist could only hope to minimize the risk of being taken unawares.

A lot of thought was given to deciding which organisms might serve as
the most reliable 'indicators' of what was really happening in the wider
world (Moore 1965a). The feeding habits of the species and the ease of
collecting specimens had to be taken into account. This presupposed the
existence of extensive and detailed records on the species. Unfortunately,
this was rarely the case. Whereas fairly complete records might be available for rarities in the past, many species were thought to be so common
that few records of their presence or absence had been made. It was only
when these species began to suffer a decline that closer interest was
taken in their whereabouts—by then it was too late to establish a base-
line.

Particular concern was expressed over the fate of the smaller birds of
prey. The nineteenth century had seen the disappearance of the large
raptors from most of lowland Britain, and there now seemed a real
danger of the small raptors and other predatory birds meeting a similar
fate. By 1963, the British Trust for Ornithology (BTO) was convinced of
the need for a survey of small predators and crows, and it once again
nominated an officer of the Conservancy as the organizer of the survey.
This time, the Conservancy was very reluctant to agree. Not only was the
officer already heavily committed, but experience with the peregrine
survey had shown that the layman found it difficult to understand why
the Conservancy's officers could be doing work which the Conservancy
wanted, but for which it was not responsible. The respective roles of the
Conservancy and BTO were all too easily confused.[4]

It was not until early 1964 that these misgivings were overcome, and
Ian Prestt was appointed director of the BTO survey in a personal capacity. A year previously, Mr Prestt had been transferred to the Toxic
Chemicals and Wild Life Section as the vertebrate ecologist. Questionnaires were sent to individuals, groups, and societies. A total of 141 were
completed and returned (Prestt 1965). The coverage of the data was
considered adequate for England, and sufficient for Wales and Scotland
to indicate whether trends in breeding success were similar to those of
England. The levels of accuracy were also thought to be high. The survey
was concerned only with large-scale trends over wide areas; it related
only to breeding birds; the species were comparatively easy to identify;
and the observations were generally made by experienced local recorders. Although everyone taking part in the census knew that a decline
was thought to be taking place due to toxic chemicals, the results for the
different species and regions were by no means uniform. Indeed, an
extremely varied picture emerged.

Of the species studied, all the birds of prey had decreased in population, whereas all the crows had increased. The extent and severity of the decline of the sparrowhawk (*Accipter nisus*) were outstanding. Although one of the commonest and most widely distributed of diurnal predators up to 1950, it had become rare as a breeding bird in every English county by 1963, and declines had also been reported from the remainder of the British Isles. The kestrel (*Falco tinnunculus*) had experienced a marked decline in the eastern half of England over the same period. The decline of the barn owl (*Tyto alba*) was less severe than that of the sparrowhawk, but more widespread than that of the kestrel (Prestt 1969*b*).

Most participants in the census attributed the decline mainly to the use of toxic chemicals. It was thought significant that the two raptors most severely affected, the peregrine falcon and sparrowhawk, were the principal diurnal predators for which other birds formed a major part of the diet. There was no evidence to suggest that game preservation, changes in climate or food supply, or the incidence of disease, were the primary cause of these changes in status. Whilst the buzzard (*Buteo butteo*) had been indirectly affected by the spread of myxomatosis, and the barn owl by severe winters and loss of nesting sites, the other raptors did not seem to have been similarly affected. There had, however, been serious outbreaks of the Newcastles disease, and it was known that wild birds could be infected. For six months, Ian Prestt, with scientists of the Houghton Poultry Research Station, investigated the possibility of its being a contributory factor in the decline of the birds of prey, but no support for the hypothesis was found.

In a detailed and concurrent study, the breeding distribution of the sparrowhawk was surveyed in selected areas, each known to have contained between 10 and 15 pairs before 1950. In the study area in Cambridgeshire, no pairs were found in 1964; some were found in the Suffolk area, but none bred; only one pair was found in the Yorkshire and Cumberland areas, and its eggs proved to be infertile. In Anglesey and Surrey breeding was on a reduced scale, and only in the New Forest could breeding be described as normal (Prestt 1966).

Particular interest continued to be taken in the status of the peregrine falcon. A sample census of 1965/66 covered about a third of the breeding population and provided further confirmation that the dramatic decline in population had ceased since 1963. Numbers had been relatively stable, with occupied territories at about 40 per cent of their pre-war level. The proportion of pairs that reared young successfully varied between 13 and 16 per cent of the maximum possible. The geographical pattern was unchanged, with breeding success and brood size highest in the south and east Highlands. There was some evidence of a marginal improvement in breeding success in southern Scotland and northern Ireland (Ratcliffe 1965*b*, 1967*b*, 1969).

Ten years after the original BTO survey, a complete census of the breeding peregrine was attempted in 1971, again directed by Dr Ratcliffe. In addition to the established network of peregrine-watchers, a grant from the Conservancy enabled the BTO to employ three full-time surveyors. About 90 per cent of the territories known to have been occupied at some time since 1930 were covered. Of these, 54 per cent were occupied, and 25 per cent had young, compared with 44 and 16 per cent respectively in 1963. Not only had the decline levelled off since 1963, but a slow and local recovery could be discerned. The British population was rapidly assuming considerable international importance in view of the otherwise global decline in numbers (Ratcliffe 1972).

The results appeared to provide further support for the hypothesis that the original decline in numbers and breeding success had been due to exposure to persistent pollutants. No tenable alternative hypothesis had been found. Instances of high levels of contaminants were still reported, not only from intensively cultivated areas but even from predominantly pastoral districts. An egg taken from an eyrie in Nithsdale, Dumfries-shire, in 1971 contained the highest level of dieldrin ever recorded in a British peregrine egg, namely 5.0 ppm.

The only modification of the original hypothesis was a growing realization that organochlorine pesticides might not be the only persistent chemicals affecting the status of the peregrine. Industrial pollutants might also be involved. Ratcliffe (1972) found it significant that there were still no peregrines, and consequently no breeding success, in the vicinity of the large maritime bird colonies, where large numbers of auks, kitti-wakes, and fulmars (*Fulmarus glacialis*) were usually taken, whereas some recovery was recorded where the peregrine depended mainly on inland prey, or such prey as *Larus* gulls, *Sterna* terns, rock-doves (*Columbia livia*), and coastal waders. This suggested that the peregrines of the cliff-lined, rocky sections of the coast, feeding mainly on maritime species, were much more exposed to persistent toxic chemicals derived from marine pollution. Analyses of eggs and tissues from birds in these locations revealed not only pesticide residues but also industrial and other chemicals. A bird from Dunnet Head had no less than 200 ppm of polychlorinated biphenyls (PCBs) and 54.6 ppm of organochlorine insecticide residues in its liver. Even a bird from the Norfolk coast had 17.4 ppm of PCBs. Dr Ratcliffe concluded that exposure to toxic chemicals still provided an entirely adequate explanation of enhanced mortality among the peregrine population and, where multiple contamination occurred through maritime food chains, the total load of toxic chemical pollutants could easily account for the further decline in status.

Between 1963 and 1965, specimens of a wide variety of carnivorous species were sent to the Toxic Chemicals and Wild Life Section from all parts of Britain. Rather than confuse the public with a variety of requests for field specimens, the Conservancy and the Royal Society for the Protection of Birds agreed in late 1962 that the Society would be responsible for making the appeals, mainly through its quarterly journal, *Birds*, and that the Conservancy would be informed whenever corpses were available of the species required. The results of any analyses would be given to both parties.[5]

Sparrowhawk, kestrel, tawny owl (*Strix aluco*), and barn owl made up the majority of the specimens received by the Section through the RSPB appeals. Many specimens had been found beside roads. From analyses of 459 livers and 364 eggs, mainly carried out by the Laboratory of the Government Chemist, it was clear that residues of dieldrin and pp'-DDE (a metabolite of DDT) were almost always present, and those of heptachlor epoxide were widespread. Most specimens contained, however, only relatively small amounts of particular compounds. In a review paper of 1965, Dr Moore suggested that residues of 10 ppm of dieldrin and 30 ppm of pp'-DDE could be regarded as a rough indication of a lethal concentration. (Moore 1965*d*). Table 9.1 identifies the proportion of specimens found in this category. All were taken after the ban in 1962 on the use of aldrin, dieldrin, and heptachlor as seed-dressings in spring. Most were collected after the more extensive ban on organochlorine pesticides in 1964 (Prestt 1967).

In view of the sudden, severe, and widespread decline in the peregrine and sparrowhawk populations, considerable concern was expressed for the future status of the grebe and heron. These two species had the highest levels of residue recorded in a nationwide survey of organochlorine insecticides in the breast muscle of 21 taxonomic families (Moore and Walker 1964). The BTO readily agreed to a suggestion from the Conservancy that a national survey should be organized of adult great-crested grebe (*Podiceps cristatus*) in May 1965. The population was found to be significantly higher than that recorded in a similarly thorough survey of 1931. A sample census of 1946–55 suggested that a steady increase had taken place since 1949.

In their report to the BTO, Prestt and Mills (1966) concluded that pesticides had had no marked effect on population numbers. Relatively low average liver-residues were found. The increase in population seemed to reflect the creation of additional expanses of water in the form of new reservoirs and flooded gravel pits. These observations were borne out in a further report, in which Prestt and Jefferies (1969) described a closer study of a sample of inland bodies of water.

In the same way, pesticides could not be held responsible for any obvious decline in the heron population (*Ardea cinerea*), which had the highest residues ever recorded in a British predatory species. Because of

Table 9.1 Proportion of specimens examined containing potentially lethal concentrations of pesticide residues

	Number examined	dieldrin > 10 ppm	pp'-DDE > 30.0 ppm	hept. epox. > 10.0 ppm	Combinations of at least 2 of the other 3 categories	Total
Kestrel	116	12	3	1	2	18
% of total examined		10%	3%	1%	2%	16%
Barn-Owl	75	3	1	1	1	6
% of total examined		4%	1%	1%	1%	8%
Sprarrow-Hawk	44	1	6	0	1	8
% of total examined		2%	14%		2%	18%
Tawny Owl	45	1	1	0	0	2
% of total examined		2%	2%			4%
Total	280	17	11	2	4	34
% of total examined		6%	4%	1%	1%	12%

breeding surveys conducted by the BTO since 1928, more was known about the heron population than perhaps any other bird species. The BTO decided to organize a further census in 1964, and a detailed breeding study was carried out by the Toxic Chemicals and Wild Life Section on two Lincolnshire heronries. (Milstein, Prestt and Bell 1970). The total population and overall output of young from the colonies appeared to be the same as in 1928. It was not clear whether the general level of residues in the birds was too low to have any serious sub-lethal effects, or whether those birds with high levels constituted so small a proportion of the breeding population that they had no marked effect on overall performance (Prestt 1966, 1970). There was a suspicion, however, that the slow recovery of the population after the severe winter of 1962–63 may have been due to the sub-lethal effects of residues. The heron population always drops markedly following severe winters, but on this occasion recovery took a particularly long time.

Despite obvious difficulties in interpreting the evidence, birds were extremely useful indicators of pollution or some other adverse trend in the environment. They were easily identified and the surveys had proved that comparatively accurate population counts could be made of them. In a paper in *Nature*, Moore and Tatton (1965) suggested that the eggs of seabirds might prove particularly useful indicator organisms. Most of the egg residues analysed in the surveys of 1963–64 had fallen within a much smaller range than the egg residues from terrestrial species.

Ideally, the chosen indicator species should be equally useful in an international monitoring programme, but there were inherent difficulties. Even if the species were cosmopolitan, there was no guarantee that its feeding habits would be the same throughout its range. Some species were too uncommon over much of their range to be suitable as indicator species, and others, including the starling (*Sturnus vulgaris*), contained very low levels of residue. Moore (1966*b*) concluded that the eggs of the cormorant (*Phalacrocorax carbo*), little tern (*Sterna albifrons*) and oyster catcher (*Haematopus ostralegus*) might prove the most adequate indicators of contamination on an international scale.

During a symposium on chemical pollution at the Fifteenth International Ornithological Congress in 1971, Ian Prestt and Derek Ratcliffe presented a paper reviewing the literature on the impact of organochlorine insecticides on European birdlife. Most records were derived from Britain, Norway, Portugal, the Netherlands, Sweden, and France. Residues were reported in at least 154 different species from 47 taxonomic families. This represented just over three-quarters of the families and about one-third of the species that occurred in Europe. The geographical distribution of the birds containing residues could be related directly to the use of pesticides in Europe. Chemical analyses of 132

Passeriformes, collected between 1965 and 1967 in Norway, revealed that 37 per cent contained no residues. In Britain, only 1 per cent of the 150 bodies and eggs of *Passeriformes* collected in a sample between 1963 and 1966 was free of residues. The difference was taken to reflect the more limited use of pesticides in Norway, compared with southern England where most of the British specimens were collected. Residues were, however, found in all the predatory species examined, even in remote areas. They were absent in only one of the 25 specimens included in the Norwegian survey. There were many instances of a bird containing as many as three different types of residue (Prestt and Ratcliffe 1972).

Birds were not the only taxa to be used as indicator species. The outbreak of fox-death in the East Midlands had played a major part in alerting conservationists and sportsmen to the possible dangers of seed-dressings in 1959–60 (see p. 68). Throughout the 1960s and early 1970s, Dr Jefferies was in correspondence with the Masters of Fox Hounds and carrying out autopsies on the corpses of foxes found in different parts of the countryside.

Attention was also focused on the badger (*Meles meles*), following reports of its being found among seed-dressing casualties and of a marked decline in population since seed-dressings were first used. In 1969, Dr Jefferies published details of the circumstances surrounding the death of 17 badgers in south-east England during 1964–68. The livers of 8 of the animals were analysed to detect levels of organochlorine insecticides. Half had very probably, and the other half had certainly, died from dieldrin poisoning. One of the badgers had been taken alive, and displayed such symptoms as inappetance, tameness, apparent blindness, and convulsive movement of the muscles, which were similar to those reported in foxes afflicted with fox-death. As death seemed inevitable, the animal was killed and analysis of the liver showed 29 ppm of dieldrin and 4.4 ppm of heptachlor epoxide. Five of the other specimens had similarly large quantities of dieldrin, namely, from 17 to 46 ppm, and the emptiness of the stomach and incidence of earth and leaf-mould in the mouths suggested that death had followed symptoms indicative of poisoning. Four specimens were found in or near water, suggestive of the animals being very thirsty—another highly significant symptom (Jefferies 1969).

Because of growing concern over the future of the bat population, the Toxic Chemicals and Wild Life Section also examined the amounts of insecticide carried by bats as 'background' in East Anglia and, by laboratory experiments, assessed the body-residues of DDT and metabolites after lethal poisoning. In examinations of 30 bats of five species, taken between 1963 and 1970 from around Monks Wood, it was found that all livers contained pp'-DDE, 82 per cent had pp'-DDT, and 29 per cent had dieldrin. The most likely source was the food supply of flying insects. The quantity of DDT in the bat livers examined during the flying period was at a maximum in March, after hibernation, and was lowest during November. This was attributed, firstly, to the fact that the storage fat laid down before hibernation would contain large concentrations of insecticide. As this fat became depleted during the winter, considerable amounts of DDT material would be released. Secondly, DDT useage was particularly high in the early part of the bats' flying period.

Laboratory tests with pipistrelles (*Pipistrellus pipistrellus*) indicated that they were more sensitive to DDT than any other mammal tested, except for the American big brown bat (*Eptesicus fuscus*). By comparing the residue levels recorded in the laboratory with those in specimens, Jefferies (1972) concluded that bats taken at random in the study area were carrying a third of the lethal level of DDT and DDE, with just under the lethal level being present after hibernation. Both individuals and the population were at risk; a high proportion was likely to be experiencing sub-lethal effects.

The movement of pesticides through the environment

A high priority of the Toxic Chemicals and Wild Life Section was to study the passage of persistent pesticides through different ecosystems. Because the ultimate repository of the residues of many pesticides was the soil, some of the earliest field experiments sought to discover whether residues led to any imbalance in soil fauna. On the Castor Hanglands National Nature Reserve, Dr Brian Davis found that the millipede, *Glomeris marginata,* was eliminated from grassland plots for six years by a single application of dieldrin (5.4 kg/ha) (Davis 1964*a*; Moore 1967*c*).

Although a variety of organochlorine insecticides had been found in ground-feeding birds, very little was known about how they reached the birds. As a first step, Dr Davis examined the feeding preferences of a range of bird species found on different crops growing around Monks Wood (Davis 1967). An assessment was also made of the likelihood of soil animals acting as vectors of the organochlorine insecticides (Davis 1966).

Because so little was known about levels of insecticide present in those invertebrates that formed the food of common bird species, Davis and Harrison (1966) collected 10 soil and 10 invertebrate samples, representative of conditions in Huntingdonshire. The dieldrin and pp'-DDT residues in worm samples were 2 to 10 times higher than those in the soils of the same sites, and were generally higher than in corresponding beetle samples. The residues of pp'-DDT, pp'-DDE, and pp'-TDE in soils from orchards were not only greater than those from arable sites, but were largely concentrated in the top part of the profile (Davis 1968). Further trials established that higher residues were picked up by invertebrates feeding on the soil surface after a foliage application of pesticides than by subterranean species after a soil application (Davis and French 1969). From these various studies, there seemed every indication that soil invertebrates could provide a regular source of pesticide residues for ground-feeding birds.

The opportunity to investigate the passage of pesticides through the forest ecosystem arose in 1963 as a result of an invitation from the Forestry Commission to collaborate in the surveying and monitoring of the effects of spraying against the pine looper moth (*Bupalus piniaria*) in Cannock Chase.[6] The collaborative exercise highlighted the persistence of pp'-DDT in forest soils and pine needles, and the considerable amounts of pp'-DDE and pp'-TDE accumulated in birdlife after spraying. Before the 1963 programme, the forest and ride-side soils still contained 8.7 and 16.0 per cent respectively of the levels of pp'-DDT recorded after an earlier spraying programme of 1954. Between 5 and 6 weeks after the 1963 spraying programme, 32 and 37 per cent respectively of the dosage was recovered. Residue levels in such species as the coal tit (*Parus ater*) were monitored. The average residues of pp'-DDE in 4 liver samples taken one week after spraying were as high as 10.9 ppm compared with a level of 0.37 ppm recorded in 3 samples before treatment. Levels of 1.98 ppm were recorded in 5 specimens a year later (Walker 1966).

How were these residues transported around and stored within organisms? In a paper in the *Journal of Wildlife Management,* Jefferies and Davis (1968) described an experiment whereby the soil was treated with dieldrin and the passage of dieldrin from soil to earthworms to song thrushes (*Turdus ericetorum*) was recorded. At the highest dose rate of 381 micrograms (μg) per day contained in worms fed to the birds, 8.5 per cent of the dieldrin consumed was accumulated by the bird, 84.2 per cent was metabolized, and 7.2 per cent was excreted as dieldrin. The bird experiencing this dose rate died in 9 days. The remaining 4 birds were killed and analysed at the end of a 6-week period. A comparison of the residue levels recorded in the birds fed on a diet ranging from 0.32 to 5.69 ppm of dieldrin in worms produced results that were to have a considerable bearing on future diagnoses of the causes of death. If the residues found in the surviving birds were extrapolated, they suggested that the bird on the highest dose rate should have had, after 6 weeks, 7.77 ppm in the liver and 3.12 ppm in the brain. In fact, it had 17.94 ppm and 16.88 ppm respectively at the time of its death on the ninth day.

On the basis of these findings, Jefferies and Davis (1968) set out the hypothesis that, at some critical point, poisoning could trigger off a reduction in the fat reserves of the body. This would then release greater quantities of dieldrin into the blood, and thus produce a rapid rise in the concentrations in the brain and other organs just before death. The bird could have drawn on these fat reserves either because it had ceased to eat or because the hyperactive stage of poisoning consumed considerable amounts of energy.

Whatever the precise reasons for the mobilization of dieldrin previously 'stored' in the fat reserves of the bird, it was clear that any diagnosis based only on dieldrin levels in the brain or some other organ of a live bird might be very misleading. In the experiment, the bird fed on the second highest dose rate of 174 µg/day was in much greater danger than would previously have been supposed from the levels occurring in its liver and brain. Its total body-burden of pesticides was actually greater than that of the bird that died and could have become lethal if its fat reserves had been metabolized (or mobilized). This could happen not only as a result of the effects of the pesticide but also from stress brought about by migration or the winter season.

The importance of metabolic changes was emphasized in feeding trials conducted in the laboratory on a caged bird, the Bengalese finch (*Lonchura striata*). Jefferies and Walker (1966) described how pp'-DDT was mixed with a caged bird rearing-food in concentrations of between 75 and 1,200 ppm by weight. Analyses of the livers of two dosed birds, extracted within 10 minutes of death, indicated that the principal metabolite of pp'-DDT was pp'-DDE in the living avian liver, and that pp'-TDE was formed as a result of post-mortem reduction. This was later confirmed by studies made on livers from wood-pigeons.

There was a strong suspicion that levels of residues, considered sub-lethal in laboratory trials, could become lethal under field conditions. Dr Jefferies found that a kestrel, maintained in a cage while being experimentally fed with dieldrin, required considerably more to kill it than two kestrels fed on a similar diet but allowed to fly. Only the caged bird was able to accumulate large fat reserves during the trial (Prestt and Ratcliffe 1972).

The Toxic Chemicals and Wild Life Section was particularly keen to examine the effects of stress on the peregrine falcon. Because of the practical difficulties of using wild specimens, the opportunity was taken to analyse the corpses of two lanners, (*Falco biarmicus*), which had been trained for falconry and had died for no obvious reason. From these analyses, and those of four peregrine falcons taken in Britain and one in Holland, Jefferies and Prestt (1966) speculated that the initial decrease in peregrine population had been caused, not so much by the gradual accumulation of toxic chemicals, but by a rapid build-up. The birds had eaten a comparatively small number of pigeons, containing high levels of residue derived from feeding on dressed seed. Support for this hypothesis of a quick build-up of residues was obtained from post-mortem examinations carried out on four rough-legged buzzards (*Buteo lagopus*) (Prestt, Jefferies and Macdonald 1968).

During studies of the incidence of toxic chemicals in different species, an anomaly had been noted. Although the long-tailed field mouse (*Apodemus sylvaticus*), bank vole (*Clethrionomys glareolus*) and short-tailed vole (*Microtus agrestis*) might carry very small residues of organochlorine insecticides, the birds which fed on these mammals frequently had high, and at times lethal, quantities of dieldrin. Either the kestrel, barn owl, and other predators differed from other species in being able to accumulate most of the dieldrin ingested, building up these residues over a longer period of time, or the pesticide content of the mammals fluctuated greatly in the course of a year, with the highest levels after the dressed seed had been sown.

The apparent anomaly was investigated in a series of trials. In a small-scale study of November 1966, the mean total body loads of dieldrin were recorded in small mammals trapped during a 17-day period before drilling, and a 13-day period afterwards. All the *Apodemus* fed on the dressed grain as soon as it was sown. The whole-body loads rose 68 times, from 2.46 μg to 166.8 μg of dieldrin. Concentrations increased during the two weeks after drilling, and caused mortality in some specimens (Jefferies, Stainsby and French 1973). A further trial in 1973 revealed that, even after two months, body levels were still appreciably higher than those recorded before drilling had taken place (Jefferies and French 1976).

The implications of these findings were assessed on the basis of a study in 1969, when bank voles were fed for 24 hours on wheat dressed with dieldrin at field-rates, and subsequently killed and fed to laboratory-reared kestrels. It took the equivalent of only seven voles, fed at these rates, to kill a kestrel in seven days. It was also noted that the abnormal behaviour of these highly-dosed mammals might attract the attention of a predator. Several times, the voles were seen running in tight circles away from cover, sometimes backwards, and often emitting loud squeaks.

On the basis of these observations, Jefferies *et al.* (1973) drew attention to the speed with which a pesticide, even when used correctly as to the season, dosage, and crop, could move into the environment. They also pointed out a deficiency in the voluntary ban that had been imposed on the use of dressed wheat. Because it was thought that seed-eating birds ate little grain in autumn as opposed to spring, the ban had not been applied to autumn dressings. As a result, the partial ban was only of limited value to those predators that took small mammals. In order to protect these birds, a complete ban on the use of organochlorines as seed-dressings was required.

Not only did the various studies instigated by the Toxic Chemicals and Wild Life Section highlight the variety and complexity of physiological processes that had to be taken into account, but they emphasized the need to measure and understand even more precisely the relationship between exposure to contamination and levels of residue in the organisms affected. Unless these essential pre-requisites were met, it was not possible to predict the likely consequences to wildlife of different degrees of contamination. There would be no possibility of using wildlife species as an indicator of the degree of environmental contamination (Moriarty 1972a).

If these requirements could be met, there would be the opportunity of studying the relationship of exposure to accumulation and, in particular, of looking at the concept that exposure to accumulation might be influenced by the position of an organism in the food chain, or web. Because carnivores ate herbivores, and were in turn eaten by other predators, it was easy to assume that persistent pesticides could accumulate along the food chain. Indeed some observers went so far as to assume that, if 10 herbivores ate and retained equal amounts of a persistent substance, the carnivore that ate these individuals would contain 10 times as much of the substance as each of the herbivores consumed.

In examining this concept, the first step was to discover whether the major source of contamination was the food of the organism or its physical environment. By the late 1960s, a number of experiments had confirmed that fish could absorb dieldrin directly from water (Chadwick and Brockson 1969). Whilst the highest levels of pesticide residue in terrestrial species were found in those at the end of food chains, namely predatory birds and mammals, field and laboratory studies cast doubt on whether this was entirely due to the accumulation of residues along the food chains. Whereas levels were high in the heron and great-crested grebe, they were much lower, for example, in two other flesh-eating species, namely the goosander (*Mergus merganser*) and red-breasted merganser (*M. serratus*) (Walker and Mills 1965). Where predatory species had high residue levels, Moriarty (1972a) suggested that this was the result not so much of their position in the food chain but of their comparative inefficiency in disposing of foreign compounds. There might also be a tendency for individuals of these species to select prey with relatively high, but not lethal, levels of residue in their bodies. The behaviour arising from these sub-lethal levels made prey much more vulnerable to predation.

In their trials with song thrushes (*Turdus ericetorum*), Jefferies and Davis (1968) discovered that, far from the levels of residue increasing as one progressed from the soil to the earthworm to the thrush, there were only 18.4 to 24.9 ppm of dieldrin in the earthworms that had spent 20 days in soils containing levels of 25 ppm of dieldrin. The levels recorded in the earthworms were much higher than the total body residues of 0.09 to 4.03 ppm found in the birds. Not only was there a decrease in levels at each step in the chain, but the role of metabolism and excretion in removing residues was highlighted by the trials. Only 0.8 to 4.4 per cent of the ingested insecticide had been retained by the birds.

According to Moriarty (1972c), the misconception as to how pesticides accumulated in food chains had arisen from a failure to distinguish between the degree to which pesticides might remain in the physical environment and in individual organisms. Whereas half the application of an organochlorine insecticide might remain in the soil for over ten years in some instances, there were no documented cases of such persistence in a living body. Metabolism and excretion made it possible for animals to remove foreign substances from their bodies.

A 'compartmental model' had long been used by scientists studying the effects of the intake and rate of loss of drugs in human beings and other organisms, and Moriarty (1975a, c) used such a model to explain what might happen to organochlorine residues. In Fig. 9.1, a compartment is defined as the quantity of DDT, or some other residue, which has distinctive and uniform kinetics of metabolism and transport to and from that compartment. Residues enter the blood (compartment 1) from the physical environment, or from food, at a steady rate (R) and they move to other compartments within the body via the blood. The rates of transfer are indicated in the model by the rate constants (K). Excretion occurs from the blood, whilst most metabolism in vertebrates takes place in the liver (compartment 2).

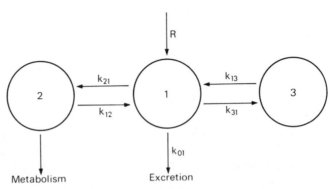

Fig. 9.1 Model for the distribution of DDT within an organism that is considered to consist of three compartments.

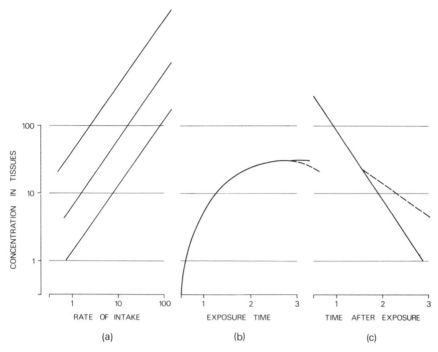

Fig. 9.2 Relationships derived from the compartmental model between exposure to pesticide and concentrations in tissue.

Whilst the movement of residues in the body was clearly much more complicated than this simple model suggested, it served to show how residues might accumulate. The level of concentration in a tissue or organ depended on the rate of intake (Fig. 9.2a), with differences of as much as a thousand-fold between the levels recorded in individual compartments. The highest concentrations usually occurred in fatty tissues, presumably because organochlorine pesticides were fat-soluble. If the rate of intake was constant, and continued long enough, concentrations within the body would eventually reach a plateau level, where excretion and/or metabolism of the pesticide took place as quickly as it was being absorbed. In some cases, and after a lapse of time, further changes in tissue concentrations might occur (Fig. 9.2b), owing perhaps to the ageing of the organism or changes in the levels of those enzymes that metabolize the pesticide. If exposure was reduced or eliminated, tissue concentrations would drop more rapidly at first than later (Fig. 9.2c). This would reflect the way in which the pesticide could be excreted directly from the blood whereas it had to pass first from other tissues into the blood before being excreted (Moriarty 1972*a*).

The value of the compartmental model was that, by estimating the rate constants, or the parameters derived from them, it was possible to measure differences between species in the way in which they took in, accumulated, distributed, and disposed of pollutants. It helped to identify more precisely the interactions of pollutants and the effects of other variables (Moriarty 1975 *b*).

The sub-lethal effects of pesticides

The widespread incidence of pesticide residues in the environment could not longer be disputed. Debate now focused on whether the levels of contamination recorded could cause death. According to Ratcliffe (1972), a lot of sterile argument had focused on the extrapolation of experimental toxicological data on captive birds to populations of wild birds. It was argued that if the residue levels in wild birds found dead in the field were below an experimental LD_{50}, there was no evidence of their having died from lethal poisoning. In Dr Ratcliffe's view, such arguments overlooked the probability of inter-specific differences in sensitivity to toxic chemicals, and they ignored the way in which wild birds were exposed to a wide range of natural stress. Death came to wild birds in many different forms, and anything that enhanced environmental rigours or impaired physiological efficiency would further shorten life expectancy.

Even where the lethal effects of a pesticide could not be proven, it was likely that individuals within the population were suffering from sub-lethal effects. In the relatively easy conditions of laboratory life, such effects might appear to be insignificant but, in the field, even a very slight impairment in efficiency, or an alteration in behaviour, might have important consequences. Dr Ash of the Game Research Association later recalled how, during the early days of seed-dressing deaths, he had noted that it was the cocks that usually died straightway, and that a second wave of mortality, principally affecting hens, occurred a month or two later during incubation. The additional stress of egg-laying and incubation, and the loss of body fat, may have tipped the scales for those birds carrying sub-lethal residues.[7]

From 1963, the Toxic Chemicals and Wild Life Section embarked on a series of experiments to evaluate the effects of residues on metabolic rates, hormonal balance, and reproductive success in a variety of species. Not only did insects provide outstanding experimental material but there was, in some cases, a direct conservation interest. Some species, and particularly a number of butterflies, were thought to be declining as a result of insecticide use. As a first step, the effect of DDT and dieldrin on the small tortoiseshell butterfly (*Aglais urticae*) was studied. The pesticides were dissolved in a volatile solvent, and drops of the solvent containing 1.25, 5.0, or 20.0 micrograms of dieldrin were put on the larval cuticle; those larvae acting as a control were dosed with solvent alone. Only the latter laid many eggs. The largest dose killed half the larvae and resulted in complete sterility of the surviving adults. Egg fecundity was affected by smaller doses than egg fertility (Moriarty 1968).

It was relatively easy to demonstrate the presence of a sub-lethal effect, but much harder to prove whether it had a significant impact on population dynamics in the field. Two questions had to be answered. At what level would the pesticide begin to exert an effect, and what would be the influence of different internal and external environments? The results of trials suggested that, outside the sprayed areas, the species was unaffected by organochlorine insecticides but, as Moriarty (1972b) stressed, this was an enormous extrapolation from a few detailed facts, taking no account of the possibility of a different result being obtained if the specimens were exposed to chronic, as opposed to single topical, doses. It also ignored the possibility of there being further variables affecting the insect's response.

The sub-lethal effects of pesticides could help to explain why high levels of residues were found in the bodies of some predatory species. It was well known that predators tended to seek out the weaker members of a prey population. If an important cause of this weakness was the sub-lethal effect of a pesticide, the predator would, in effect, be selecting out the most contaminated individuals, and thereby increasing its own exposure to the pesticide. Such a hypothesis was extremely difficult to prove under laboratory conditions. As Moriarty (1975b) related, when a house sparrow was placed in a room with butterflies, some of which had been dosed, the bird was either too full to eat anything, or it ate the whole lot so quickly that the sequence of consumption could not be recorded!

Perhaps the most satisfactory experimental situation arose out of a series of studies made by Dr Arnold Cooke on the effects of DDT on the tadpoles of the common frog (*Rana temporaria*). Because of allegations that the use of agricultural chemicals might be a contributory factor in the decline of the species, particularly in eastern England, he kept tadpoles for an hour in amphibian saline containing nominal concentrations of 0.01, 0.1, 1.0, and 10.0 ppm of pp'-DDT, and thereafter in fresh saline. Cooke (1970) noted a series of quite different behavioural changes. At first, the tadpoles entered a 'frantic' phase of tail-lashing, body-twisting, and rapid swimming, followed by a 'resigned' phase characterized by persistent slow swimming in a twisting manner. Finally, they either became moribund and died, or returned to normal activity.

As Cooke (1970) stressed, the normally high levels of mortality made it difficult to measure the implications of these various changes induced by DDT. In a series of experiments, however, Cooke (1971) introduced single warty newts (*Triturus cristatus*) into an aquarium with pairs of tadpoles, one of each pair being normal and the other hyperactive from DDT-poisoning. In 90 trials out of a 100, the dosed tadpole was eaten first, strongly suggesting that newts were more likely to prey upon hyperactive tadpoles than normal ones. This may not necessarily have been harmful to the prey population because there was evidence that contaminated individuals would have died in any case at metamorphosis, when the small residues, stored in lipid reserves, were mobilized. Predation merely removed a source of pointless competition from the healthy individuals. The effects on the predator might be much more important. An extrapolation to other predator-prey systems suggested that high levels of exposure could occur where predators susceptible to poisoning sought out relatively resistant prey.

Because most studies of the sub-lethal effects of pesticides on birdlife had used gamebirds and incubators, the Toxic Chemicals and Wild Life Section decided to use passerine birds, pairs of which would incubate and rear their own chicks. The greater stress thus placed on the birds would resemble more closely actual conditions in the field. Because British birds were difficult to breed in large numbers for laboratory use, a caged bird, the Bengalese finch (*Lonchura striata*), was used.

In a paper in *Ibis*, Jefferies (1967) described a significant correlation between the intake of DDT in female birds and delay in ovulation; he attributed this to the effects of the insecticide on the pituitary or hypothalamus. In further trials, it became clear that, although the clutch sizes of the DDT-fed and control birds were similar, the fertility, hatchability, and fledgling-success of the DDT-fed pairs were lower. There were longer periods of ovulation, a longer incubation and rearing period, lighter eggs, and the newly-hatched young were smaller (Jefferies 1971).

Since 1958, Dr Derek Ratcliffe and others had drawn particular attention to the largely unprecedented phenomenon of the eggs of the peregrine falcon, sparrowhawk, and golden eagle being broken or ejected from their nests by the parent birds (see p. 109). Could egg-breaking be another of the sub-lethal effects of pesticides, and could it have been a contributory factor in the 'crash' in population numbers? It had started well before the introduction of the dieldrin group of pesticides, but after the introduction of DDT and gamma-BHC.

The possibility of a causal link was put forward in a paper circulated among colleagues by Dr Ratcliffe in mid-1966. Responding to his tentative suggestion that decrease in eggshell thickness might be involved, Dr D. Nethersole-Thompson and Professor J. Hickey suggested that he might compare the weight and thickness of eggshells collected before and after the onset of egg-breaking, and to that end, several owners of private egg-collections were contacted. An index of eggshell thickness was devised, namely, the eggshell weight, divided by the egg-length, multiplied by the egg-breadth. In a note in *Nature* in 1967, Dr Ratcliffe illustrated that there was indeed a causal connection between the decrease in eggshell weight and an increase in egg breakage (Ratcliffe 1967*a*), and, in a paper of 1970, he described in much more detail how these were related to exposure to persistent organochlorine pesticides and a decline in breeding status.

Whereas the pre-war values of the index of eggshell thickness had been very stable, there had been marked changes since about 1946–47. The relative thickness of the eggshells of peregrine falcons and sparrowhawks had decreased by 19.1 and 17.2 per cent respectively by 1948–50 and these changes had persisted up to 1969. In the case of the golden eagle, a decrease of 9.9 per cent had begun not later than 1951 (Ratcliffe 1970). In the United States, a similar decline in eggshell thickness was reported in birds of prey, including the peregrine (Hickey and Anderson 1968).

Although DDT did not come into common agricultural use in Britain until 1948–49, military stockpiles of the insecticide were used, often on a locally lavish scale, from 1945 onwards for domestic, horticultural, and veterinary purposes. It became common, for example, to dust homing-pigeons with DDT in order to control ectoparasites, a practice which must have brought many peregrines into direct contact with the pesticide. South of the Highlands, domestic pigeons formed the principal prey-species of the peregrine. The possibility of a predator ingesting residues applied to prey in this way were confirmed many years later, when the analyst at Monks Wood, Mr M.C. French, analysed by gas-liquid chromatography a number of homing-pigeon rings, and found that one, extracted from peregrine castings and collected by Dr Ratcliffe, still bore minute traces of DDT and DDE (Ratcliffe 1980).

The degree of exposure of the peregrine to this type of contamination became clearer when the solvent rinsings of a series of peregrine egg clutches were analysed (Peakall, Reynolds, and French 1976). Those taken in 1933, 1936 and 1946 showed no traces of residues, whereas a second clutch in 1946, and further clutches in 1947 and later years, all contained traces of DDE. As Ratcliffe (1980) noted, the detection of these residues in shell membranes and egg contents, after an interval of nearly 30 years, was further proof of the persistence of the compounds.

There was also evidence of geographical parallelism. The greater decrease in weight of sparrowhawk eggshells in south-east England, compared with those from Cumberland, Hampshire, and Dorset, matched the areas where pesticides were used more intensively. In the central and eastern Highlands of Scotland, eggshell change was slight or absent in peregrines and not recorded in golden eagles, corresponding with findings that contamination of peregrines was only about a quarter of that found in the remainder of Britain, and a tenth of that found in golden eagles in the western Highlands. In the western Highlands remote from the sea, where eggshell change had occurred, it was difficult to envisage any other form of environmental change, except for the very direct one of the introduction of organochlorine sheep-dips (Ratcliffe 1970).

Dr Jefferies believed that most sub-lethal effects (as opposed to direct effects) were likely to be induced by a basic form of disturbance. If the identity of that disturbance could be discovered, it might then be possible to predict the sub-lethal effects more precisely. With this end in view, he devoted considerable attention to the role of the thyroid and metabolic rate. A 'unifying theory', whereby a close similarity was described between the observed effects of organochlorine residues, and those recorded for hyper- and hypo-thyroidism, was eventually published (Jefferies 1975).

It was found that when sub-lethal amounts of DDT were fed to domestic pigeons (*Columba livia*) during initial trials, there was an increase in the weight and activity of the thyroid and a considerable reduction in the colloid content of the follicles (Jefferies and French 1969). Doses of DDE and dieldrin had a similar effect (Jefferies and French 1972*a*). In further studies to see whether this response reflected a hyper- or hypo-thyroid state, 79 pigeons were fed control diets and dosage rates of pp'-DDT. After 6 to 11 weeks, the birds were showing hyperthyroidism at the lowest rate (3 mg/kg/day), and hypothyroidism at doses of 6 to 54 mg/kg/day (Jefferies and French 1971). It was not known how DDT produced these different responses, or why the change-over in response occurred, but Jefferies (1973) contended that, either by stimulation or competition for binding sites, the effects of these residues on the avian thyroid could constitute one of the most important forms of sub-lethal lesion.

So far, almost every experiment designed to find out why insecticides caused thin shells had involved the examination of the organs or body fluids of the affected birds. Dr Arnold Cooke decided to adopt a different approach: he studied the detailed structure of shells from a variety of sources as a means of indicating what changes in biochemical or physiological processes might have occurred in the laying bird. Cooke (1975) compared the thin shells laid by chickens fed on calcium-deficient diets and on diets containing chemicals known to interfere with calcium deposition in shells, with the thin shells laid by captive ducks treated with DDT and those shells produced by wild herons known to have been exposed to organochlorine insecticides.

   The shells of the chickens and herons proved to be broadly similar in so far as there were reductions in the thickness of the two main component layers of the shells. These were probably caused by a reduction in the availability of shell components in the shell glands of the birds. By contrast, some parts of the thin shells laid by the ducks were at least normal in thickness. This suggested that they may have resulted from premature termination of the growth of the calcified part of the shell, rather than from a 'shortage' of shell components. Cooke (1979) subsequently compared the contemporary and older shells of the sparrowhawk and peregrine. He found a roughly proportional reduction in thickness of the two main component layers (the mammillary and palisade layers), suggesting that these thin shells resulted from a decreased rate of deposition. Significantly, shell-thinning was shown to result in a proportionate reduction in shell strength.

   Whatever the reasons for thinner eggshells, the Toxic Chemicals and Wild Life Section had little doubt that they contributed to the substantial decrease in reproductive success in the raptors concerned. In the sample of peregrine eggs collected between 1963 and 1968, mean shell-index was 1.60 in those from clutches which later produced fledged young compared with 1.46 for eggs from nests which failed to produce any young. Research on other raptors both in Britain and abroad suggested, however, that any attempt to measure more precisely the relationship between shell thinning and breeding success was likely to be extremely difficult.

   From the Conservancy's Scottish headquarters, Dr Ian Newton, in collaboration with colleagues within and outside the Conservancy, compiled records for 325 nests, and for eggs from 130 clutches, of the sparrowhawk in Dumfriesshire over the period 1971–73. This was an area where the bird still bred in large numbers. Compared with 71 clutches collected in the same general area in 1900–45, those taken in 1971–73 showed an 18 per cent reduction in shell index. No chicks were hatched in a third of the nests recorded in the latter period and there was partial failure in a further third of the nests. Eggs breakages accounted for 31 per cent of all failures. Although the breakages were generally related to the degree of shell thinning, the eggs in 22 per cent of the clutches affected had normal shell indices, whereas those in 66 per cent of the clutches with low indices survived. Clearly some other factor, such as parental care, was also involved. The main cause of unsuccessful breeding was a failure to lay eggs in a nest that had been constructed. This accounted for 43 per cent of all complete failures. The third commonest cause of failure was desertion (15 per cent). The main residue found in the eggs was DDE. The organochlorine levels recorded in those clutches where breakages occurred, and in those where the eggs were incubated successfully but failed to hatch (11 per cent of complete failures), were significantly higher than in those showing normal success (Newton 1974; Newton and Bogan 1974).

In a review of the causal relationships between environmental pollution, decrease in eggshell weight, egg breaking, breeding failure, and population change, Ratcliffe (1970) argued that these relationships were likely to be multidimensional (Fig. 9.3); the different hypotheses being put forward to explain them were by no means mutually exclusive. Much more research was required. As Prestt and Ratcliffe (1972) pointed out in their review of the effects of organochlorine insecticides on European birdlife, it was only in Britain that detailed information could be found on the possible effects of the pesticide on wild populations. And even here, it related only to a few of the larger predatory species. In no country was there any detailed research to show the effects on other groups, despite the fact that large numbers of gamebirds, pigeons, and passerines were known to have died in large numbers following the use of insecticides.

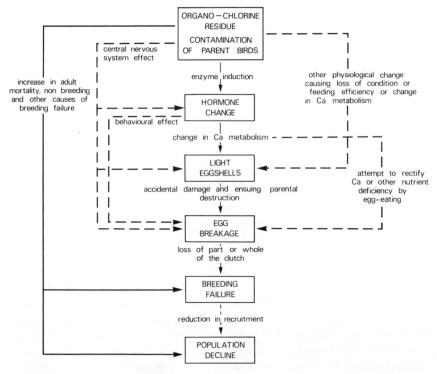

Fig. 9.3 Possible relationships in a causal chain induced by organochlorine residues.

The pesticide treadmill

The research programme of the Toxic Chemicals and Wild Life Section fell into two parts, namely, surveys of the incidence of pesticide residues in different forms of wildlife, and studies of the implications of their presence for the survival of these species and of wildlife communities in the future. It was the farmer who first used the term 'pesticide treadmill' to describe how, once used, pesticides had to be applied in ever-increasing quantities thereafter. The term was soon taken up, and the literature abounded in instances where a pesticide, designed to control one pest species, upset the natural controls regulating the population of another species, which subsequently assumed the status of a pest. There were also instances of a resurgence of the original pest later in the season, owing to the elimination of its natural predators.

At a Discussion Meeting of the Royal Society, Dr Moore commented that 'we are so used to thinking in terms of "a pesticide destroying a pest and having side effects" that we tend to forget the essential nature of the reaction'. Although a pesticide might be directed against one or two specific species, it was in fact, he said, applied against the whole ecosystem. Even if pesticides had been entirely specific (which they were not), their effects would still be far-reaching. Dr Moore drew an analogy with the most outstanding form of biological control in modern times, the spread of myxomatosis among rabbits in the mid-1950s. Not only had the disease led to the virtual demise of the rabbit, but the status of many other species had changed as a result of the almost total removal of these grazing animals. If this could happen when the 'pesticide' was specific, 'we must not be surprised if the application of a pesticide which affects many species has radical effects on the ecosystem to which it is applied' (Moore 1967a).

Natural ecosystems, Moore said, were so complex that no one had unravelled the normal workings of even one of them and, like other ecologists, those studying the effects of pesticides could do little more than construct very crude models, reflecting the complexity of reality (Moore 1965d). As a prerequisite to building these models, the ecology of such features as the hedgerow in the agrarian landscape was investigated. Dr E. Pollard assessed the effects of hedgerow management on animal populations, as exemplified by small-mammal populations (Pollard and Relton 1970) and the hoverfly predators of the cabbage aphid, *Brivicoryne brassicae* (Pollard 1971). Another member of the Section, Dr M.D. Hooper, investigated the degree to which the origins and age of the hedgerow habitat were reflected in its present-day composition (Pollard, Hooper and Moore 1974).

In a study of the effects of controlling the small white butterfly (*Pieris rapae*) with DDT, Dr J.P. Dempster demonstrated how the interdependence of populations of different species made it extremely difficult to predict the overall effect of a pollutant on individual species. His study was one of a number to illustrate how populations of some species might 'crash', whilst others might show a remarkable increase. The precise effect depended upon the sensitivity of the individual species and of the other species that influenced its numbers. As a first step to assessing the effects of controlling *Pieris*, which was a serious pest in brassica crops, a study was made of the incidence of natural mortality. From trials conducted between 1964 and 1966, it was found that over 90 per cent of the young stages died between the egg and pupation, largely as a result of predation by arthropods. About 12 species fed on *Pieris*, the most important being the ground beetle (*Harpalus rufipes*) and the harvestman (*Phalangium opilio*) (Dempster 1967).

With this background information, Dr Dempster compared the survival rates of *Pieris* in plots where no spraying took place with those sprayed twice with DDT at a time when the number of first and second generation caterpillars was at its maximum. Although DDT was very effective in killing most of the caterpillars before they had an opportunity to damage the crop, infestation could be as bad as ever within five weeks. Butterflies from outside the treated area would continue to lay eggs on the plants, which would meanwhile produce new leaves, free from DDT. More significantly, spraying had killed most of the arthropod predators of *Pieris*, particularly those ground-living species which climbed the plants at night to feed on the young caterpillars. DDT might persist in the soil for long periods and thereby minimize, or preclude, the chances of recolonization (Dempster 1968*a*, *b*, and *c*).

A deeper understanding of population ecology suggested that, far from being a desirable attribute, the persistent effects of DDT in the soil could be a considerable disadvantage in controlling *Pieris* because of the way in which they led to a reduction in *Harpalus* and other predators. It was a further example of the dangers of relying solely on a particular type of compound and form of pest control (Dempster 1975).

Instead of applying ever-increasing quantities of pesticide to combat the progressive loss of natural predators, alternative ways of reducing pest populations had to be found. In a world-wide review of 30 years of experience in crop protection based on chemical insecticides, Metcalf (1980) cited the increasing number of studies suggesting that the only means of escaping the pesticide treadmill was to develop the concept of integrated pest management (IPM), whereby the disadvantages of the use of pesticides were minimized and the maximum use was made of the benefits conferred. IPM focused on the total well-being of the crop and the ecosystem in which it grew.

'The further review'

Whilst it had a distinctive research programme, the Toxic Chemicals and Wild Life Section represented only a very small part of the research effort being invested in the study of the effects of pesticide use in Britain and overseas. In a review paper on eggshell thinning, published in 1973, Dr Arnold Cooke was able to cite an enormous volume of literature that had appeared since the first reports of the phenomenon in 1967 (Cooke 1973). Not surprisingly, there was fierce debate on the interpretation of the research data from different laboratories, and on the significance of the conclusions reached in the light of the benefits otherwise conferred on farming by pesticide use. How successful were the Conservancy's scientists in fending off criticism and promoting their findings? Perhaps the most effective way of answering the question is to describe how the Conservancy continued to participate in the controversy as to whether further restrictions on pesticide use were necessary.

There was little point in being aware of what was happening in the natural environment unless there was scope for influencing the course of events. In a leader in the RSPB magazine, *Birds,* Stanley Cramp warned that Britain was in danger of falling behind other countries in regulating the use of pesticides (Cramp 1969). In March 1969, Sweden imposed a two-year ban on the use of DDT and was soon followed by Denmark, Hungary, Norway, Italy, and a number of American states. The United States Forestry Service announced that chlordane would replace the more persistent pesticides used for the control of various forest-invertebrate pests.

The Ministry of Agriculture and other parties on the Advisory Committee on Pesticides struck a cautious note. In reply to a Parliamentary Question, a spokesman for the Ministry of Agriculture explained that the stricter control of DDT was one of the questions being examined by the Advisory Committee. It would be wrong to follow blindly the lead of other countries. In his words, 'each country is different from the other'. In a Supplementary Question, Lord Chorley agreed, but pointed out that 'the human body is the same'.[8]

The Nature Conservancy also struck a 'cool, factual note'. A delicate situation had arisen from the recent restrictions imposed on the use of cyclamates in human foodstuffs, following an American inititative. Whilst some commentators were quick to draw encouragement from this onslaught on cyclamates, Gerald Leach may have spoken for many more when he stated in the *Observer* that the main lesson of the cyclamates issue was the need for more discrimination in the control of pollutants. 'Instead of acting emotionally, or bending under the blast from every quick slogan or fervent lobby group, we have to realize that these things are never as simple as they seem—and think about them hard before we move'.[9] A further constraint on the Conservancy in making any pronouncement was the imminent publication of *The further review of certain persistent organochlorine pesticides used in Great Britain* by the Advisory Committee on Pesticides.

As promised in its report of 1964, the Advisory Committee had embarked on a *Further review* in 1967, under its new chairman, Sir Andrew Wilson. In submitting evidence to the Committee, the Conservancy described how the research of the Toxic Chemicals and Wild Life Section had been geared to provide as much information as possible in readiness for the *Further review*. The evidence suggested that, whilst the voluntary bans had made the recovery of some species, possible, the rates of recovery were slow. Not only had there been a continuing loss of adult raptors, but widespread breeding-failure. The Conservancy again called for a complete ban on aldrin and dieldrin as seed-dressings, and on all other uses of the compounds for a period of three years. There should be a detailed assessment of the agricultural value of every persistent organochlorine insecticide, which should, in any case, be phased out as soon as adequate alternatives could be found.[10]

The stature of the Conservancy as a scientific adviser had risen considerably since 1963–64. At that time, the Conservancy had not only drawn attention to the widespread presence of these compounds in the environment, but had been the first to highlight the attendant dangers of allowing them to accumulate. The warnings had soon been endorsed by reports of persistent residues being found in other localities, remote from where the pesticides had been applied. The extent of contamination was perhaps most graphically underlined by the discovery of residues in the wildlife of Antarctica, first at the US Antarctic base at McMurdo Sound (George and Frear 1965), and then at the British Antarctic Survey base in the South Orkneys (Tatton and Ruzicka 1967).

This confirmation of the Conservancy's warnings did not protect its scientists from being assailed by a barrage of criticism. As expected, agricultural and medical experts argued that the value of the insecticides in food production and preventative medicine more than outweighed any risks to the environment and wildlife. In its evidence to the Advisory Committee, the National Farmers Union spoke of the need to safeguard the farming industry, reduce food imports, and compensate for losses arising from the conversion of 30 000 acres of farmland to urban development each year. Far from helping farmers to meet these challenges, the banning of certain uses of pesticides would lead to higher costs. It was grossly unfair to expect British farmers to compete with overseas producers, who remained free of such controls over their use of toxic chemicals.[11]

Not surprisingly, the scientists employed by the pesticides industry subjected the findings of the Conservancy's scientists to the closest scrutiny. Thus Shell Research Ltd, using data from its own laboratories, promoted a vigorous campaign defending the use of dieldrin. A paper published by scientists of the Shell Tunstall Laboratory described feeding trials using Japanese quail (*Coturnix coturnix japonica*) and domestic pigeons. The concentration of dieldrin in the brain after death was found to be independent of the dosage, the period that had elapsed before death, and loss of weight before death. The concentration in the liver was also independent of dose and loss of weight; there was, however, a tendency for the concentration to increase as the time of death approached (Robinson, Brown, Richardson and Roberts 1967).

A paper, first given at a conference on toxicity in the United States in 1968, denied that there was any evidence to suggest 'the existence of a new and generally inimical factor' affecting bird species in Britain since the introduction of organochlorine insecticides. Robinson (1969) cited the Common Bird Census, which, he claimed, had shown no serious effects attributable to these insecticides in the period 1962 to 1966. The alleged significance of a relationship between the decline of sparrow-hawks and peregrines, and the use of the insecticides, could not be 'tested rigorously as the published information is too imprecise'. Furthermore the sub-lethal effects of organochlorine residues did not appear to be significant in the most common bird species, and their effects on raptors since 1962 'appeared rather tenuous'.

Such assertions, where supported by data, deserved respect and consideration. What the Conservancy found so disconcerting was the bitterness and niggling opposition that emanated from those scientists and administrators who had previously denied any possibility of pesticides having harmful side-effects, and who now seemed so resentful at being proved wrong that they would stop at nothing in fighting a rearguard action. It was not an easy time for those in the Toxic Chemicals and Wild Life Section.

Much of the criticism was made 'in confidence', or off-the-record, but Ratcliffe (1980) recalled how:

some of us had our first experience of scientists playing politics, and we learned how vicious a vested interest under pressure can be. It was clearly in many people's interests, one way or another, to believe that the wildlife conservationists were talking nonsense, and they left no stone unturned in trying to establish this. Every new paper with more evidence was dissected and gone over with a fine tooth comb, to see what flaws could be found. Some of the toughest opposition came, not surprisingly, from the agrochemical industry's own scientists, but certain members of the Governments agricultural establishment were well to the fore. Tactics at times resembled those of the courtroom rather than the scientific debating chamber. There were tedious arguments about the nature of proof and the validity of circumstantial evidence. The attempts to deny effects of pesticides on wild raptors descended now and then to obscurantism.

Some insight into the mutual antagonism between scientists may be gained from remarks made in a presidential address to the Association of Applied Biologists by Dr D.L. Gunn, a former Director of the International Red Locust Control Service in central Africa. As scientific adviser to the Secretary of the Agricultural Research Council, he also served as a member of the Advisory Committee on Pesticides. In his presidential address of 1971, Dr Gunn spoke on the theme of 'Dilemmas in conservation for applied biologists'. Extreme conservationists, he said, had attacked the work of these biologists without inhibition, and often without much regard for truth. Their expositions on the impact of pesticides had often been one-sided with invective, and the use of the smear technique had been commonplace. Because biologists involved in improving modern technology, public health, and agriculture had made so little response, Dr Gunn contended that public opinion had been 'mainly formed by deliberate propaganda and by what readers of newspapers and books (and watchers of television) find exciting and not merely dull and true' (Gunn 1972).

In setting out to expose the falsehoods perpetrated in the name of conservation, Dr Gunn cited the Conservancy's correlation of eggshell thinning with the introduction of DDT. It was another example, he said, of such conservation bodies having offended 'objective judgement'. It was as if conservationists had found a structure with some good angles, but some loose ends (Fig 9.4a) and, in order to tie the loose ends together, had created another structure (Fig. 9.4b). In going on to attack the Conservancy's ill-formed structure, Dr Gunn tended to be much more dogmatic than Dr Ratcliffe had ever been in his papers. Yet in dismissing the hypothesis of a casual link between eggshell thinning and the use of DDT as 'utterly unacceptable', Dr Gunn may have overlooked the fact that he had himself also failed to find detailed information on the amounts of DDT being used at different dates, and the precise chronology of eggshell thinning.

Throughout the discussions of the late 1960s on the implications of eggshell thinning, Dr Moore insisted that the fundamental argument did not hinge on whether there was a precise correlation between the two sets of data. The essential point was that the coincidence appeared significant when the lack of change in eggshell thickness over the previous 50 years was taken into account. Obviously, some of the pieces of information had greater scientific validity than others. Knowing, however, that DDT could affect calcium metabolism, that experimental work at Monks Wood and independently in the United States had shown a close correlation between eggshell thinning and DDT content, and that reductions in

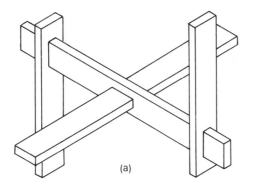

(a)

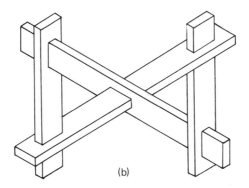

(b)

Fig. 9.4 Good and bad structures.

reproduction could be correlated with the presence of residues, Dr Moore argued that it would have been thoroughly irresponsible for the Conservancy not to have brought the hypothesis of a casual relationshp to the attention of the Advisory Committee on Pesticides.[12]

The wide divergences in the interpretation of scientific data may have been reflected in the long delay in publishing the *Further review.* Promised in 1967, the report was two years late by the time it was published in December 1969. For many, it was the apotheosis of moderation and circumspection—a masterpiece of draftsmanship rather than a programme for action. Critics claimed it represented, not so much the collective wisdom of the members of the Advisory Committee, but the deep divisions that separated them. So wide was the gulf in views that, in the words of Jon Tinker in the *New Scientist,* 'this committee was hard put to agree on anything other than platitudes'. The report was so skilfully worded that if the chapter headings and recommendations were removed, the remainder could equally well serve as a demand for the total and immediate ban on all organochlorine pesticides or for government aid in promoting that part of the pesticide industry (Anon 1969; Tinker 1970).

For the Conservancy, the drafting of the *Further review* was a difficult time. At least five drafts were prepared. At one of the drafting stages, Dr Moore recalled that the two overriding objectives of the report were to reassure the public that there were no acute dangers to man resulting from organochlorine use and to warn the public that the theoretical reasons against the use of organochlorine insecticides had now been reinforced by recent discoveries in the laboratory and field. Whilst the latest draft succeeded in meeting the first objective, the Conservancy believed insufficient weight had been given to the second and it feared this would weaken the whole report in the eyes of the conservation movement. Although the report and its component parts were designed to represent the collective view of the Advisory Committee, the conservation movement was bound to attribute those passages dealing with wildlife to the Conservancy's representatives. It would expect detailed references to be made to the sub-lethal effects of organochlorine insecticides, and particularly to those on eggshells.[13]

In its report, the Advisory Committee noted the growing concern over the general rise in environmental pollution resulting from new technological processes and the disposal of waste products. Particular attention had been paid to pesticides because their effects could be identified far more easily than those of many other forms of pollution. Although organochlorines had been singled out for particular criticism, the Advisory Committee believed it would be wrong to attach any great priority to their withdrawal. There were far more dangerous forms of pollution to be tackled first.

There was no evidence to suggest, the report continued, that the contamination of the environment by organochlorine pesticides was becoming worse. The discovery of residues almost anywhere in the world reflected the fact that scientists now had the skills and knowledge to make world-wide studies possible. Over the previous few years, the amounts present in the human body and food, and in wildlife, might have actually declined, whilst the populations of peregrine falcons, sparrowhawks, kestrels, and golden eagles had increased in some areas, following the restrictions placed on the use of organochlorine pesticides in 1962 and 1964.

The report denied that there was any evidence of the existing levels of exposure being harmful to man. Although most biologists agreed, many were surprised that the Advisory Committee chose to be so dogmatic, particularly in view of the misgivings recently expressed by such a reputable body as the National Cancer Institute in America. Turning to wildlife, the report conceded that some casualties were still reported where DDT was used on an extensive scale, but the levels of residue found in natural waters continued to be well below those likely to affect the survival of fish. Although DDT was very persistent in soils, the residue levels, even where application was made regularly, seldom exceeded the equivalent of a single treatment.

The paragraph of the report dealing with eggshell thickness had been redrafted so many times that it allegedly delayed the completion of the report by six weeks (Anon. 1969). It recounted that there was evidence of a reduced thickness of eggshells, increased incidence of egg breaking by parent birds, and a reduction in breeding success (though not necessarily sufficient to affect population size) among predatory birds. There was a moderately close correspondence in time between these phenomena and the increased usage of DDT, but this evidence was not precise enough to establish a casual relationship with certainty. Other factors which had not been investigated so fully might be relevant or even significant.

Although there was no cause for alarm, the report insisted that there was no room for complacency. It was 'undesirable that the human environment should contain substances capable of producing toxic effects and whose continued presence conveys no benefits to human survival and well-being'. The overall use of organochlorine compounds should be reduced and alternatives adopted, where these would not interfere with food production or cause undue hardship to farmers and other users. To that end, the Advisory Committee proposed further restrictions (Advisory Committee 1969).

In a statement to parliament in December 1969, the Minister of Agriculture endorsed the Committee's findings, and expressed the Government's hope that, by the end of 1970, the sale of DDT in small packs to home-gardeners would have ceased and, likewise, that the use of aldrin, dieldrin, and DDT on certain crops, and of dieldrin and DDT for certain purposes in food storage and in the household, would also have ended. The Minister affirmed the need to keep all the remaining uses of pesticides under close scrutiny, with a view to progressively adopting less persistent forms as soon as they became available.[14]

For its part, the Conservancy considered issuing a statement welcoming the report and amplifying the evidence on the sub-lethal effects of pesticides and wildlife. The statement was drafted, but never used. There was concern lest its publication might be interpreted in some quarters as an attempt to publish a minority report.[15]

It was easy for the nature conservation movement to be critical of the halting steps being taken towards the withdrawal of these pesticides but, for the more seasoned observers in that movement, the Committee's report and the Minister's announcement represented a major advance in the protection of wildlife. There was at least a commitment to phase out DDT and the other organochlorine compounds. Not only would the commitment be of direct benefit to wildlife, but it could be taken as an implicit endorsement of the advice given, and the research pursued, over the previous decade.

PESTICIDE USE:
THE OPTIMAL FORM OF CONTROL

The safe use of pesticides depended on their being used responsibly and on knowing how they might affect the areas being treated. The first called for education, and the second for more research. Opinion differed on how far the Conservancy should be involved in both the educational *and* research aspects. Compared with the pesticide and farming industries, the Conservancy's resources were extremely limited.

In order to place the later discussions on the Conservancy's advisory and educational roles in perspective, it is necessary to retrace events since the late 1950s. The Conservancy's dilemma was given prominence in July 1961, when an Adjournment Debate was initiated in the House of Commons on the research role of the Conservancy. In a largely hostile speech, Marcus Kimball attacked it for failing to give adequate guidance on the effects of toxic chemicals on wildlife. Far from the Conservancy taking its place as a third advisory service on the countryside, together with the National Agricultural Advisory Service and Forestry Commission, he said, there had been 'a sad lack of contact between those responsible for the conservation of nature and those who are likely to damage or destroy it'. In responding to the Debate, the Parliamentary Secretary for Science, Denzil Freeth, stressed that the Conservancy was a research council, and therefore lacked both the organization and resources to conduct a publicity or educational campaign.[1]

Yet lack of money and manpower could not absolve the Conservancy entirely from entering the 'public-relations' field. Not only did it want to secure active support for its work in protecting wildlife and the natural environment, but, as a national body receiving grant aid from the government, it had a duty to explain its role and point of view to the general public. In an office note of 1958, Mr Boote had stressed the urgency and enormity of the task. Despite widespread interest in nature subjects and the success of such BBC programmes as 'The naturalist', 'Birds in Britain', and 'Look', only 'a minute fraction of the public' had a clear idea of what conservation and ecology meant.[2] The designation of nature reserves and promotion of research, he said, were not enough—an educational role had to be developed as fast as resources would permit.

One of the main objectives of the Monks Wood Experimental Station was to provide field facilities for college and university courses. Under the aegis of the Conservancy, an informal Study Group on Education and Field Biology was convened and it published a report on 'Science out of doors' in 1963 (Study Group 1963). In conjunction with the Council for Nature, residential courses were also held for teachers As Mr Boote, one of the convenors, commented, they were valuable opportunities for conservationists to discover the best ways of disseminating information.

The Conservancy was alive to the need for closer links with the mass media, the various land owner and user groups, and the public at large. The pesticides issue provided further insights into and experience of ways of 'getting the message across'. Whatever forms of pesticide control were adopted, whether mandatory or voluntary, none could succeed until knowledge replaced ignorance. To be competent in the use of pesticides and other aids to farming, users had to understand the full implications of what they were doing

The dissemination of advice

Whatever its wider role in public relations, the Conservancy had a statutory duty to advise those whose everyday activities brought them into contact with wildlife and the natural environment. How was it possible for the Conservancy to steer a course between the expectations of many and the constraints laid down by reality?

During 1960–61, Dr Moore spent much time trying to identify those points where liaison should be fostered and developed. Following a symposium on 'Insecticides, fungicides and the soil', he reported that 'the scientists present understood the extreme complexity of the ecological problems resulting from the use of toxic chemicals', but the manufacturers who employed them had still not faced up to the implications of this complexity. Whilst very willing to collaborate and to take the advice of scientists, they did so in the hope that simple solutions might be found to difficult problems. Through closer liaison, the Conservancy had a role to play in removing these false hopes.[3]

It was the task of the Toxic Chemicals and Wild Life Section to make abundantly clear how pesticides were affecting wildlife, and how the harm might be allayed. In 1963, Dr Moore addressed the annual conference of the British Association (Moore 1964b) and the Section mounted an exhibition for the Royal Society's Conversazione at Burlington House. In 1964, a BBC film on the effects of toxic chemicals included an interview with Dr Moore and film showing the work at Monks Wood. In addition to their research papers published in academic journals, members of the Section wrote articles for magazines with large circulations, and those directed at the farming and pesticide industries. The Conservancy was also under pressure to issue some kind of official code or statement. A leaflet, *Pesticides and wildlife,* was eventually distributed with the Annual Report of 1963.[4]

Two publications stand out from all the rest. The first was, as its title suggests, 'A synopsis of the pesticide problem'. It was written by Dr Moore for the fourth volume of the series, *Advances in Ecological Research.* As the editor of the series remarked, much had already been written on pesticides, but mainly in terms of specific uses or compounds. Dr Moore broke new ground by perceiving pesticides as a new ecological factor that would affect total ecosystems. For its holistic approach, the review marked an important stage in the pioneering development of pesticide/wildlife research. It set out an agenda for further research that would aim to close the many gaps in knowledge that remained (Moore 1967c).

But by far the most widely-read publication was Kenneth Mellanby's book, *Pesticides and pollution,* published as the fiftieth volume in the famous 'Collins New Naturalist Library'. In praising the style and scope of the book, the chairman of the Conservancy, Lord Hurcomb, 'urged that the Conservancy's Senior Officers be encouraged and given time' to write similar works on other subjects.[5] Completed in March 1966, the book attempted to provide an objective account of pollution, and 'to fit pesticides into the general picture of pollution from all sources' (Mellanby 1967). In its first 10 years, 15 000 hardback and 60 000 softback (Fontana) copies of the book were sold and there was a separate Book Club edition.

Both in terms of its style and presentation, the appearance of the book was well-timed. The layman needed a review of the information-explosion that had occurred in the previous few years. During that time, Dr Mellanby had been Director of the Monks Wood Experimental Station and a member, *inter alia,* of the (Frazer) Research Committee on Toxic Chemicals. In the words of a reviewer in *Nature,* he had 'obviously used the past few years very valuably indeed', and had produced the informative, comprehensive, and balanced account required. The *Sydney Morning Herald* reviewed the book in the context of the growing concern in Australia about the effects of pesticides. Although written for the British reader, the implications arising from the book's message were international in significance. Perhaps the strength of the book was most succinctly summed up in a leader in *Amateur Gardening,* which observed:

The great thing about this book is that it has no axe to grind: it bows neither to manufacturer, farmer, nor naturalist, but shows the necessities as well as the dangers of our present situation, as well as examining every chemical in use.

The most refreshing aspect of the book, and the key to its considerable success, was that, by the last chapter, the reader had ceased to be interested in whether the author was for or against the forces which led to pesticides and pollution. The author's overriding concern was that society, and the individuals who made up society, should know what they were doing—in other words, that they should be competent in their use and management of the world in which they lived. No one could quarrel with those sentiments.

It was essential that the interest engendered by articles and books should be followed up by personal contact. As early as the winter of 1961–62, Dr Moore had given 15 lectures on conservation and the work of his Section. In a paper to the Conservancy's Scientific Policy Committee, he set out his impressions. Very often, he wrote the misuse of chemicals would be avoided if farmers felt that, by conserving wildlife, they were contributing something useful to public welfare. It would be a great help if such bodies as the Ministry of Agriculture and National Farmers Union published 'definite public statements, commending the conservation of wildlife on farmland'.[6]

Mutual respect is not achieved by drawing up headings of agreements, convening formal meetings, and issuing joint-communiqués. The 'main business' is usually done elsewhere. A sense of rapport develops out of a mixture of informal and formal contacts that often pass unrecorded and with little publicity. Those who recall the tentative moves towards the closer regulation of pesticides in the early 1960s stress the role of the 'Wig and Pen' luncheons, attended by individuals from industry and the conservation movement. These meetings left no formal record, but their significance was none the less for that. Their bearing on the events recorded in the files should never be overlooked.

A significant initiative was taken in 1963, when a liaison committee was set up between the Conservancy, voluntary bodies, and the Association of British Manufacturers of Agricultural Chemicals (ABMAC). Reporting on its first meeting, Mr Boote commented on the remarkably wide range of agreement that had emerged. For its part, the Conservancy was keen to discover precisely what steps were being taken by manufacturers to safeguard wildlife. What instructions did the individual firms give their sales representatives?[7]

One of the more tangible results of this new liaison was the symposium on 'Agricultural chemicals—progress in safer use'. Organized by ABMAC, and chaired by the chairman of the Conservancy, the symposium was held in March 1964, with an opening address by Lord Zuckerman. The proceedings were reported by the Agricultural Correspondant of *The Times* under a heading, 'Call for more education in the use of pesticides'. A leader in the same edition warned that some ecological changes were bound to follow the otherwise beneficial use of pesticides, but 'it ought to be shown in advance what these changes are likely to be'.[8]

Both ABMAC and the Nature Conservancy laid great emphasis on involving European manufacturers in these deliberations, especially in view of the multi-national character of many of the leading companies and the international traffic in pesticides. In submissions to the Expert Committee for the Conservation of Nature and Landscape, which had been appointed by the Council of Europe (see page 140), Mr Boote emphasized the need for conservationists to win the respect and under-standing of those whose activities would bring about large-scale changes in the environment. To be successful, conservationists had to take every opportunity of pressing and testing their arguments. Only in that way could 'the fullest range of common objectives' be agreed.[9]

In May 1964, Mr Boote addressed a meeting of European manufactur-ers of pesticides, and he later recorded:

I am personally satisfied that there is every opportunity for co-operation between conservationists and chemical manufacturers and that, with goodwill on both sides, it should be possible to get the same level of co-operation that we have in this country.

It was highly relevant that these initiatives came at a time when confer-ences were being organized in Britain with the aim of bringing every conceivable type of organization together to discuss the common theme of 'the Countryside in 1970'. The concept of a European Conservation Year was being developed. Far from being some peripheral activity, the skills and experience being acquired as part of the study and regulation of pesticides were of direct relevance to the fast-developing conservation movement by the mid-1960s.

In late 1961, the Conservancy and Ministry of Agriculture discussed ways of improving collaboration between the two bodies.[10] The Conser-vancy identified three important areas for closer liaison, namely:

(1) co-operation and mutual support in selected research projects;
(2) development of ways and means of bringing appropriate conser-vation principles and practices into the normal pattern of farming;
(3) assessment of the impact of recent, or new, developments in land use.

When staff were canvassed on ways of achieving closer collaboration, a number of practical difficulties soon emerged. With existing staff alloc-ations, it was hard enough to carry out existing procedures, let alone expand them. The Toxic Chemicals and Wild Life Section devoted about 40 days a year to committee work. There was a 'very grave danger' of members becoming so diverted by committee and other liaison work that they would have little time left for the research on which their advice would ultimately be based.[11]

The Ministry of Agriculture issued leaflets and press notices on a wide variety of topics. A notice of 1961 may be cited, calling for 'greater co-operation and consultation between beekeepers, farmers, fruit growers, market gardeners and spraying contractors'. Bees were good friends of the farmer and fruit grower and, as well as producing honey, they were essential for pollinating fruit and seed crops. The press notice set out six elementary precautions to be taken when using insecticides and weed-killers.

The Ministry also tried to reduce contamination of soils and bodies of water caused by the dumping of waste products and containers. Farmers were advised to wash out containers before the spraying tank was completely filled so that the rinsings could be emptied into the tank. Experiments indicated that a single washing would remove 95 per cent, and two washings 99 per cent of the insecticide. Following a meeting with interested parties in March 1965, the Ministry and the British Crop Protection Council agreed on a 'Code of practice for the disposal of unwanted pesticides and containers on farms and holdings'. It was the subject of a succession of press notices.[12]

In all these initiatives, the Conservancy was cast in the role of adviser, and often only one of many advisers, to another body responsible for drawing up and implementing a code of practice, which was usually voluntary and left to the goodwill of others to obey and enforce.[13] It was bound to be a frustrating exercise for the Conservancy. A deft blend of patience, tact, and discretion was required, which hardly conformed with the role of a strident, assertive pace-maker, expected of the Conservancy by the voluntary conservation movement.

The non-agricultural use of pesticides

The conservation movement had long argued that the terms of reference of the Advisory Committee and the Pesticides Safety Precautions Scheme should be extended to cover non-agricultural situations. Not only were toxic chemicals being used extensively for controlling vegetation along rivers, canals, railways, and roadside verges, but it was felt that a code of practice should be applied to the industrial and domestic uses of these chemicals.[14] In 1964, 50 tons of dieldrin were used in the manufacture of woollen articles: 4 tons more than were used in agriculture.

Once again, it took a series of crises to focus attention on a subject which had previously been dismissed as too abstract or unimportant. In September 1963, a hundred cats and dogs died in Merthyr Tydfil after having eaten the flesh of a pony sold for pet food. Meanwhile, a number of dogs died in Smarden, Kent, after having eaten the offal from some 20 cows that had died in mysterious circumstances. The only link between the two incidents was the circumstantial evidence that the animals might have been poisoned by a pesticide containing fluoroacetamide.

The Government was quick to emphasize that the deaths did not arise from the recommended use of pesticides in agriculture or elsewhere. The press and some members of both Houses of Parliament argued that the incidents nevertheless highlighted the need for closer surveillance. Although not directly involved in either incident, the Conservancy took the keenest interest in the circumstances in which the domestic and farm animals died, and the response made by the authorities. Mr Boote recalled how he had repeatedly stressed the inadequacy of testing products within the 'purely scientific conditions' of the laboratory. Not only should the products be tested under field conditions, but the Advisory Committee, in making its assessments, should anticipate carelessness and abuse in the real-life world in which the compounds were used.[15]

The incidents demonstrated how easily lethal chemicals could escape into the environment, and how difficult and expensive it was to remedy the situation. It was discovered that the pony had been handed to the local knackery at Merthyr Tydfil by the police, who found it dead on a rubbish heap. When some of the flesh was fed to a dog, the Ministry of Agriculture's Veterinary Laboratory realized for the first time how susceptible dogs could be to secondary poisoning by organo-flourine compounds. In that sense, the incident provided the Laboratory with a scientific breakthrough, but a report to the Minister emphasized that there was no way of telling how the pony came across the poison. Circumstantial evidence suggested that it was in the form of rat bait, but the lack of certainty contributed to further public unease. What was to prevent children being the next victims?

The Ministry of Agriculture was extremely worried lest the incident should provoke a witch-hunt against all pesticides. The Advisory Committee had given its provisional approval to fluoroacetamide in the period before the Conservancy was represented on the Scientific sub-committee. Under the Agricultural Chemicals Approval Scheme, it was used for controlling aphids on sugar beet and brassicas, and as a rodenticide for controlling rats in sewers, ships' holds, and other confined areas. At the request of the Minister, the Advisory Committee made a fresh assessment, and advised that it should no longer be used in food production and storage. The advice was accepted, and a circular was sent to local authorities, stressing the need for rodenticides to be used with the utmost care.[16]

Critics argued that the use of fluoroacetamide should have been banned completely. In an Adjournment Debate which he promoted in February 1964, John Farr warned that the compound could still escape into the environment, even from sewers. He drew attention to the way compost fertilizer was sold by sewage farms. In reply, the Joint Parliamentary Secretary of the Ministry of Agriculture emphasized that fluoroacetamide was the first poison to clear rats from entire sewer-systems. For the first time, there was a real possibility of clearing whole towns of rats, and the Ministry was not prepared to sacrifice such a great benefit to public health for the unproven benefits to be derived from extending the already considerable restrictions.[17]

In the same month, the Minster of Agriculture told parliament that his officers were called to the Smarden area of Kent by local veterinary surgeons in April 1963. Scientist discovered that certain fields, ditches, and ponds were contaminated by an organo-fluorine compound, derived from sludge dumped at the rear of a factory manufacturing a number of chemical preparations.[18] What the Minister's statement did not emphasize was the considerable confusion and delay that had occurred before the cause of resulting deaths was ascertained.

A local veterinary surgeon first became involved in January 1963 when two puppies, belonging to an employee at the factory, had died. The owner suspected poisoning but, because no specific poison was suspected, a post-mortem examination was considered pointless. When the vet was called to some sheep on a farm about a mile from the factory, he learned of other animals dying in mysterious circumstances. The stream flowing through the fields adjacent to the factory was cloudy, and the vegetation black and dead. In May, a goat died, and its owner, another employee in the factory, revealed that fluoroacetamide was manufactured there, among other pesticides. By this time, samples of water and the internal organs from the dead sheep had been analysed by the County Analytical Laboratory. Although fluoride was present to the extent of 5 ppm in the water, there were a hundred times that amount of bromide. The watercourses were fenced off, and the levels of bromide were soon back to normal.[19]

That was not the end of the matter. A month later, a fox hound on a nearby farm went mad. Another farmer complained of falling milk yields, and some of his calves, born strong and healthy, died within a few hours of convulsions. To everyone's surprise, a team from an Oxford laboratory found flourides in the water at concentrations of 3 to 23 ppm. Although this was well above normal levels in that part of Kent, there was little concern until it was realized that the compound was an organo-flourine. As the veterinary surgeon discovered in his local library, the action of this compound was to interfere with the supply of nourishment to the brain and other organs. It was the almost perfect poison, being inconspicuous in odour and taste, very stable, and having no known antidote. The delayed action, absence of unique pathological symptoms, and the great difficulty in identifying residues in the body, made it especially difficult to track down. At Smarden, it killed not only farm and domestic animals, but there was circumstantial evidence of wild birds and rabbits being affected. 'Silent spring' had become a reality in that part of 'the Garden of England'.

The Nature Conservancy described the vet's unpublished report of December 1963 as being so important that, unless its lessons were thoroughly applied, further incidents would occur. Some compounds were so lethal that they should either be banned, or supplied on prescription to licensed operators. The incident reinforced the Conservancy's view that the Advisory Committee should be given wider terms of reference and made completely independent of the Ministry of Agriculture or any other interested party. Manufacturers should be compelled to carry out ecological tests at the formulation stage of a new product. Only in these ways could adequate safeguards be built into the surveillance of pesticide use.[20]

The Ministry of Agriculture insisted that the trouble arose entirely from an industrial accident, when exceptional concentrations of the poison escaped onto neighbouring farmland. Accidents were likely to occur in any highly-industrialized nation, despite all the precautions taken. In the House of Lords, however, Lord Douglas of Barloch challenged the use of the term 'accident'. It had been the company's practice to wash the silt, containing the poison, into a yard where it was possible for residues to become dissolved in water and flow into adjacent watercourses from whence they passed through the soil into vegetation, later eaten by livestock. The risks and consequences of this happening should have been recognized long ago.

In July 1963, the factory owners, Rentokil Ltd, stopped producing fluoroacetamide preparations at Smarden and arrangements were eventually made for dumping all the factory sludge, ditch water, and soil from neighbouring fields off the Continental Shelf, under Government supervision. These arrangements led to a further bout of anxiety. The British Trawlers' Federation expressed alarm over the possible contamination of fish stocks. The Joint Parliamentary Secretary, however, emphasized that the Government had taken expert advice; the canisters would be dropped into waters $2\frac{1}{2}$ miles deep.

The handling of the two incidents, involving the rubbish dump in South Wales and the dumping of contaminated water and waste from Smarden into the sea, led to a Motion being tabled by Lord Douglas of Barloch in February 1964, asking for information on what research was being conducted on the disposal of toxic wastes, and whether statutory controls should be introduced.[21] There were three other manufacturers of fluoroacetamide in Britain, and many other toxic substances were produced up and down the country. There was at present no statutory control over the disposal of their waste products.

How many more Smardens, he asked, were in the making? At

Smarden, the main remedy had been to transport the contaminated water and waste out to sea. What assurance was there that the area of contamination would not now be extended to sea and marine life? No one could be certain of this and, worse still, no one was likely to ever know because the effects of fluoroacetamide were so difficult to identify positively from all the other possible influences on human and animal life.

This time, it was the Joint Parliamentary Secretary of the Ministry of Housing and Local Government who replied to the Motion. After reviewing the work of such bodies as the Alkali Inspectorate and Water Pollution Research Laboratory, he argued that the first step was to improve understanding of how wastes should be treated. He conceded that:

we do not at present know enough about chemical industrial wastes or what are good and bad ways of disposing of them. We require, in the first place, a comprehensive list of chemicals which may become industrial wastes. We need to decide which of these are potential dangers and how they can safely be disposed of.

To that end, the Government had decided to appoint a committee of enquiry.

The incidents also stimulated direct action on the regulation of pesticides. In its report of February 1964, the Advisory Committee had endorsed the need to regulate the non-agricultural use of pesticides. An obvious course of action was to widen the terms of reference and membership of the Committee so as to embrace *all* toxic chemicals, but this was soon rejected as impractical. There were thousands of poisonous substances in use. The chairman of the Advisory Committee, Sir James Cook, argued that it would be much more practical to extend the terms of reference to allow surveillance of *all* forms of pesticide, wherever and whatever their use. Following a representative meeting convened by the Office of the Minister for Science, it was agreed that the Committee's new terms of reference should be to keep under review all risks that may arise from the use of:

(1) pesticides;
(2) potentially toxic chemicals on sale to farmers for veterinary medicines prescribed for use by veterinary surgeons;
(3) any other potentially toxic chemical referred to the Committee by Ministers, and to make recommendations to the Ministers concerned.

These proposals were accepted by the Government, and the new 'Advisory Committee on Pesticides and Other Toxic Chemicals' became responsible to the Secretary of State for Education and Science. The scope of the Pesticides Safety Precautions Scheme remained unaltered.

The Conservancy had participated in these discussions, and welcomed the widening of the Committee's terms of reference.The partial ban on organochlorine pesticides, and the enlargement of responsibilities, would put Britain ahead of almost every other country in regulating toxic chemicals. It would mean, however, even heavier demands on the Conservancy for administrative and scientific support. These repercussions were a further indication of how the question of toxic chemicals, and ultimately of environmental contamination, had become a major component of the Conservancy's work.[22]

The value of statutory controls

Despite the support given to the voluntary agreements, the Conservancy believed that some form of statutory control should be imposed on the use of pesticides. A proposal to promote a Private Members Bill in 1962 drew attention, however, to the practical difficulties that would be encountered in drafting and implementing such control. Because the Government was extremely unlikely to agree to a measure covering every aspect of pesticide use, any Bill should focus on the most essential aspects of pesticide control. Even then, there would be formidable difficulties in drafting regulations to cover particular compounds, the licensing of operatives, and the disposal of containers. In the event, the Bill was never introduced.[23]

In his statement on the report of the Advisory Committee in March 1964, the Minister of Agriculture recognized that the voluntary scheme might become inadequate 'as scientific knowledge increases and more restrictions are found to be necessary'. In a letter to the chairman of the Advisory Committee, the Minister wrote that 'the time has come when it would to appropriate to ask your Committee to review the system of controlling the use of toxic chemicals' in the light of experience. The Committee might investigate such questions as whether:

(1) approval of new chemicals should be confined to those which would be safer and more efficient than those in existing use;
(2) approval should only be given on a provisional basis;
(3) residue tolerances should be prescribed.

The Government would then consider whether the present, voluntary scheme should be put on a statutory basis.[24]

The new review came at an important juncture in the history of pesticide use. An experimental situation had been created. If seed-dressings and sheep-dips had caused losses in wildlife, as alleged, the effectiveness of the partial, voluntary ban since 1964 could be measured by the decline in residue levels and the recovery in wildlife. By 1967, 'a clear drop in the average (dieldrin) residue levels' had been recorded in mutton fat residues (Egan 1967), and gas-liquid chromatography indicated a decline in dieldrin levels in eagles' eggs from 0.86 ppm (1963–65) to 0.32 ppm (1966–67). The proportion of eyries in west Scotland successfully rearing young had increased from 31 per cent to 69 per cent in the period 1966 to 1968 (Lockie, Ratcliffe and Balharry 1969).

Meanwhile, the first steps were taken towards statutory control. Mrs Joyce Butler informed the Conservancy in June 1964 that she intended to introduce a Private Member's Bill, requiring manufacturers to identify the active ingredients in their products and state whether they were harmless to wildlife. The Conservancy warned Mrs Butler that the second requirement was 'fraught with complexity'. Although welcoming the Bill in principle, some officers of the Conservancy considered it to be piecemeal and ill-timed. The measure was largely concerned with plants, and excluded any provisions to cover the use of the products in food storage or industry, or against animal pests. It might prejudice any chances of implementing the wider measures expected to be recommended by the Advisory Committee, following its review.[25]

By her persistence, Mrs Butler succeeded in bringing the Farm and Garden Chemicals Bill to the statute book in 1967. Ministers could introduce regulations requiring pesticide products to be labelled with the name of any specified active ingredient, and for the label to bear a symbol or colour to denote and explain any hazards to humans and other forms of life arising from their use in agriculture or gardening. The first set of regulations was announced in August 1968 and, by 1971, 290 chemicals had been scheduled.[26]

The Advisory Committee confessed that it was not easy to suggest radical ways of improving the existing system, which was 'an outstanding example of voluntary co-operation between government and industry'. As government spokesmen stressed, in reply to a succession of Parliamentary Questions on the progress of the inquiry, there were many complex factors to be taken into account.[27] It was not until January 1967 that the *Review of the present safety arrangements for the use of toxic chemicals in agriculture and food storage* was published (Advisory Committee 1967).

The Committee received a wide variety of advice, but few witnesses regarded the existing system as being entirely satisfactory. No matter how well-organized and loyally supported, a voluntary scheme could never be sufficiently comprehensive to guarantee that a hazard would never occur. As the Committee argued, in its report, only a mandatory scheme could provide this greater degree of assurance. Under existing arrangements, there was, furthermore, nothing to prevent someone from importing for his own use products that had not been cleared under the Precautions Scheme. Considerable concern was expressed at the persistent failure of some manufacturers to identify on their products' labels the active constituents and the degree of their toxicity. There were some 70 chemicals, introduced before the inception of the Notification and Precautions Schemes, which still awaited scrutiny by the Advisory Committee. Whilst progress was being made, manufacturers had little incentive to go to the trouble and cost of 'clearing' those products which appeared to be not only profitable but also 'safe'. The Advisory Committee concluded that these several deficiencies could only be overcome if a mandatory scheme was introduced. Not only would it ensure that all sectors of the industry were treated equitably, but it would help to improve public confidence in the Government's ability to respond rapidly and effectively to any unforeseen difficulties that might arise in the use of pesticides and veterinary products.

A great deal of discussion centred on the concept of tolerance limits. No one denied the value in certain circumstances of prescribing limits to the amount of pesticide residue released into the environment, but the Conservancy, together with the farming and pesticide industries, was very worried lest too much significance might be attached to the tolerance level decided upon. There was a danger of farmers assuming that any residues below the tolerance level were perfectly safe. From a scientific point of view, the public might attach far too much significance to the tolerance limit. As Dr Moore remarked, in a discussion of 1966:

from my experience I believe that the public does not fully realise the there is bound to be a theoretical risk whenever a new pesticide is used, but that in the opinion of most scientists the risk is justifiable. It is odd that people do not understand the situation for it is exactly the same as the one all motorists find themselves in every day; we all know that driving a car involves some risk but we accept it as being justifiable.[28]

The lack of detailed knowledge as to what levels of which compounds had a deleterious effect on mankind and wildlife in different circumstances meant that it was possible to postpone the introduction of statutory tolerance limits indefinitely.

In order to achieve an efficient statutory scheme, while retaining as far as possible the flexibility of the existing voluntary schemes, the Committee proposed a licensing system, whereby it would become an offence to sell, supply, or import any pesticide or veterinary product for use in agriculture, home-gardens, or food storage, which had not been licensed by the appropriate ministers, in consultation with the Advisory Committee. The licensing authority would enforce the provisions with respect to the composition of pesticides and veterinary products, and the local authorities would enforce those related to such matters as labelling, advertising, and the type of containers used.

The Ministry of Agriculture accepted the Committee's recommendations in principle, and inter-departmental discussions began as to what might be included in a Bill covering pesticide use in agriculture, home-gardening, food storage, forestry, and weed control. A list of draft proposals in February 1968 identified the need to control the supply and labelling of pesticide products used in agriculture and food storage through a mandatory licensing system; the extension of provisions for wearing protective clothing when using pesticides; the designation of certain forms of misuse as an offence; the imposition of residue limits in foodstuffs; and the laying down of procedures for record keeping, warning notices, the storage of pesticides, and reporting of incidents. Such a measure would replace the Agriculture (Poisonous Substances) Act of 1952 (see p. 22) and the Farm and Garden Chemicals Act of 1967.

The proposals for the replacement of the voluntary Pesticides Safety Precautions Scheme with a statutory system were circulated to over a hundred interested parties in August 1968. For its part, the Conservancy welcomed the proposals on the grounds that they would provide, for the first time, 'a sound framework for the wider use of pesticides based on more effective controls and a closer integration of the lessons from research with policy-making and procedures'.[29] Coming so soon after a decision by the United States Federal Drug Administration to reduce the level of DDT in foodstuffs, the scientific journal, *Nature,* saw the proposed Bill as a further expression of the growing concern of governments about the use of pesticides.[30]

The major point of contention was whether the proposed Bill should cover those uses of pesticides outside agriculture and food storage. As the Chairman of the Advisory Committee, Sir Andrew Wilson, explained to the Secretary of State for Education and Science, the Committee had been given the responsibility in 1964 for reviewing the risks incurred from *all* types of pesticide use, but the voluntary Pesticides Safety Precautions Scheme had not been extended to cover uses outside agriculture and food storage. This meant that the Committee lacked information on the scale and nature of pesticide use in such places as the home, industry, or on rubbish tips, and was therefore in no position to decide whether further regulations were required in these situations. During discussions with the 'other users', the Committee had found much that was encouraging, but it held to the view that the proposed Pesticide Bill should include powers enabling the provisions of the measure to be extended, as and when necessary, to any pesticide product used outside agriculture and food storage.[31]

The Nature Conservancy attached very considerable significance to these wider powers of surveillance and control, but was anxious that nothing be done to delay or prejudice the chances of the remainder of the proposed Bill being introduced and passed. With considerable relief, it learned in June 1969 that the Secretary of State and the Minister of Agriculture had agreed to the extension of the Bill so as to include powers whereby the Government could introduce Orders for the control of pesticide use outside agriculture and food storage, should the Advisory Committee find this to be necessary.[32]

In spite of the lack of parliamentary time, and the many questions of detail which still had to be resolved, it was widely felt that the increasing attention being given to environmental pollution by Parliament and the press made legislation imperative. Drafts of Cabinet papers were prepared, seeking approval in principle for the introduction of a Pesticides Bill and guidance as to which of two kinds of Bill might be adopted. A short enabling Bill might be introduced, giving ministers wide powers to license pesticide products and to impose regulations on the supply, use, and disposal of pesticides in the light of further survey and research. It was likely that such a measure would be attacked for confering unacceptably wide discretionary powers on ministers. The alternative was a more detailed Bill. This would take much longer to draft and would be criticized by those who realized that, once enacted, there was no scope for extending the controls. Such a Bill would almost certainly be attacked by wildlife and health-food interests for not going far enough.[33]

However, despite the apparent trend towards statutory controls, a Pesticides Bill was never introduced. Not only was there a change of Government in 1970, but circumstances also changed. The pretext for a reappraisal of pesticide controls followed the publication of the first report of the Royal Commission on Environmental Pollution in 1971. Appointed in 1970 (see p. 2), the Standing Commission had identified pesticides as one of the forms of pollution that was seriously affecting the environment, and it supported the Advisory Committee in arguing that 'mandatory control is desirable and will in the end be inevitable'. There was 'already enough evidence to enable the Government to reach a decision and to introduce legislation at an early date' (Royal Commission 1971). Confronted with such powerful advocacy of the views it had expressed in its report of 1967, the Advisory Committee felt it prudent to reassess the reasons why it had advocated mandatory controls.

On reappraisal the Advisory Committee found those reasons less pressing in 1971. Experience had shown that every manufacturer, with few minor exceptions, had notified the Committee of its products, or had withdrawn them. Although it was possible to import pesticides for one's own use without prior safety-clearance, the instances of this loophole being exploited were so few and unimportant as to make legislation unnecessary—even if such legislation could have been enforced. Again, although the back log in the review of chemicals had been serious, sufficient progress had been made to leave only 20 awaiting scrutiny. There had, likewise, been a marked improvement in the standard of labelling of products, and the recent Farm and Garden Chemicals Act was likely to lead to further improvements.[34]

In a letter to the Minister of Agriculture in September 1971, the chairman of the Advisory Committee conceded that there were 'some blemishes' that could be removed 'more speedily if statutory powers were available', but the weaknesses in the 'effective and well-tried' voluntary system no longer justified its replacement. Whilst it could be argued that, in a matter of such obvious public importance, the Government should not have to rely on 'a system of control whose implementation in large measures depended upon the co-operation of manufacturers and importers with a financial interest', these misgivings were theoretical rather than practical in character. It was unlikely that the pesticides industry would ever run the risk of flouting a ministerial decision on any important issue of public and environmental safety. The recent wholehearted co-operation of industry in the phased withdrawal of certain uses of persistent organochlorine insecticides, even though this conflicted with the commercial interests of that industry, encouraged the Advisory Committee to take that view.

This change of heart on the part of the Committee left unresolved the question of how the pesticides used outside agriculture and food storage should be regulated. Even here, however, the situation had improved. There had been a suggestion from the British Wood Preserving Association that a voluntary safety scheme be established to cover that sector of use, and the Advisory Committee, in a letter to the Secretary of State for Education and Science, suggested that similar schemes might be devised for other types of use not already subject to the voluntary scheme. There was no *prima facie* evidence to indicate why these schemes should not be as succesful as the existing Pesticides Safety Precautions Scheme.

In April 1972, the Minister of Agriculture announced that legislation on the use of pesticides was no longer contemplated.[35] In a 'progress' report of 1974, the Royal Commission on Environmental Pollution agreed that there was no case for replacing the voluntary system in view of the evident co-operation of the agrochemical industry and the continued decline in the use of organochlorine compounds (Royal Commission 1974). The Commissioners argued, however, that 'the ultimate sanction' of mandatory controls should never be ruled out. It should be reconsidered in the event of any deterioration in the level of control, future technological developments, any obligations incurred as a result of Britain's admission to the European Community, or the identification of any new uses for pesticides.

At the request of the Secretary of State for the Environment, a study of the non-agricultural uses of pesticides was made by the Central Unit on Environmental Pollution. The bulk of the data was collected by February

1973, and a report was published a year later. Whilst no evidence of serious abuse emerged, the Central Unit recommended that such uses should be subject to the same degree of supervision and control as those in agriculture. The Secretary of State, Mr Crossland, accepted the recommendation (Department of the Environment 1974*b*).

The complete ban on organochlorine seed-dressings

Now that it had been decided to retain the voluntary system of pesticide regulation, the scope for imposing further controls over the use of organochlorine seed-dressings under the Pesticides Safety Precautions Scheme had once again become relevant. In late 1971, the DDT Advisory Committee of the US Environmental Protection Agency proposed further withdrawals from use of DDT. This gave rise to considerable controversy. At a public hearing and later at a press conference, Norman E. Borlaug, Nobel Peace Prize Winner in 1970 and 'Father of the Green Revolution', vociferously defended DDT for its unique contribution to the relief of human suffering. Because DDT was a name known to almost everyone, he said, it had been singled out by environmentalists. Once they had succeeded in eliminating DDT, there would be a domino effect. The search for ways of increasing food production, and of feeding the world's starving millions, would be nullified (Anon. 1971*a*; Borlaug 1972).

Borlaug's defence of DDT found support in the leader columns of *Nature*. Whilst there was a strong case for regulating the use of pesticides, there was a risk of throwing out the baby with the bathwater. Despite a decade of research on the side-effects of DDT, the leader writer contended, there was only enough evidence to warrant an agnostic position. In view of the role of DDT in food production, it was essential that 'the present wave of sentiment in favour of a ban on the use of DDT and such materials should be replaced by a proper and necessary case for regulation and supervision of the ways in which these pesticides are used' (Anon. 1971*b*).

The Conservancy would have found it easier to sympathize with these views if more positive steps had been taken to anticipate and assess the implications of trends in pesticide use. Dr Ratcliffe observed that, 'whilst the right noises were being made by all concerned, there was no commensurate activity and things were tending to drift'. Whilst the fault lay on all sides, the Conservancy could not help wondering whether its earlier success in reducing 'the number of extreme statements being made by the voluntary bodies' had led manufacturers and others to believe that the pressure was off.[36]

The Ministry of Agriculture responded to criticisms by asking for substantiated cases where losses in wildlife could be attributed to defects in the Pesticides Safety Precautions Scheme. The Conservancy accepted that there had been improvements on the detailed aspects of pesticide control, but to dwell on these was to miss the point of the Conservancy's misgivings. It was much more a matter of attitudes. The long delays in withdrawing pesticides known to be harmful to wildlife were causing the voluntary bodies to argue that the conservation movement had 'gone along' too much with industry. As Mr Boote warned in June 1972, 'if we should return to the position of the late '50s and early '60s, all our efforts will be put at risk, and the pioneering work of co-operation between government, industry and voluntary bodies of world significance could be jeopardised'.

These marked differences in attitude continued to be highlighted by the use of organochlorine seed-dressings for the control of wheat bulb fly. Each year, the agricultural departments reminded farmers that, following the voluntary agreement of 1961, those seed-dressings containing aldrin, dieldrin, and heptachlor could only be used up to 31 December. There were many complaints from the voluntary conservation bodies of farmers failing to honour the letter, let alone the spirit, of the voluntary agreements. On investigation, it was usually found that the risks to wildlife had been taken into account, but not sufficiently to satisfy the conservation bodies.[37]

The Conservancy constantly reminded the Ministry of Agriculture of the recommendation in the *Further review* that aldrin, dieldrin and heptachlor should be completely banned as dressings at the earliest opportunity. This continued to be desirable, both as a practical contribution to nature conservation, and as a gesture to the conservation movement. Spokesmen for the farming industry continued to insist that autumn dressings caused little harm to birds and that, given the need for dressings, there was no point in withdrawing organochlorine dressings until it was certain that alternatives were available and safer for wildlife. The Conservancy believed the only way of stimulating the search for these substitutes was to set a time limit on the use of organochlorine dressings but, its critics argued, this could lead to an acutely embarrassing situation if the scientists failed to come up with alternatives. The minister would have to backtrack on a firm statement that the dressings had to be withdrawn.[38]

The Pest Infestation Control Laboratory of the Ministry of Agriculture agreed with the Conservancy that the levels of dieldrin, and the number of incidents, had not fallen to the extent expected following the partial ban of 1963. Indeed, there had been 'a remarkable consistency' in the proportion of incidents caused by dieldrin over the period 1963–73. A comparison of the number of reports received each month revealed two

DIELDRIN   INCIDENTS   INVESTIGATED   1967 – 1972

Fig. 10.1 Dieldrin incidents investigated by the Infestation Control Laboratory, 1967–73.

peaks (Fig. 10.1), and proved beyond doubt that some farmers were continuing to use organochlorine spring dressings despite the ban. The losses to wildlife were, however, small. In the words of the Laboratory, they were 'a continuing but not necessarily unacceptable risk to wildlife', especially as the wood-pigeon was the most affected species (van den Heuvel 1975).

The Toxic Chemicals and Wild Life Section concluded that the only way of overcoming the impasse was to delineate areas where the damage caused by wheat bulb fly warranted the use of organochlorine dressings, and to discover whether farmers adhered to the advice given them by the Ministry. There were suspicions not only that large quantities of dressed seed were left on the surface when sowing conditions were difficult, but that the seed was often used as a poison bait against wood-pigeons and other species.[39] In the event, only a small-scale study was possible. This was, however, enough to highlight the benefits of a light harrowing immediately after drilling. Counts of exposed wheat grains had been made on seven fields near Monks Wood after autumn drillings in 1966; a further series of counts was made at the same time of year in 1972. High counts were recorded on clay soils after heavy rain, after shallow drilling on a fen field intended to reduce the risk of attack from wheat bulb fly, and after harrowing on another fen field where the farmer was attempting to bring buried potatoes to the surface. The levels of seed-dressing on the grain were highest in the first few days of exposure. Over 90 per cent of the unburied grain had disappeared within one or two weeks of drilling (Davis 1974).

Once again, the shock effect of a series of well publicized incidents brought the question of seed-dressings to the fore. In mid-December 1971, a major kill was reported from Aberlady in the Lothians, which involved not only hundreds of wood- and feral pigeons, but also geese, game birds, a kestrel and a barn owl. The birds had consumed winter corn, used in accordance with the voluntary code of practice. Publicity arising from the incident brought to light earlier incidents in lowland Scotland; there were further major incidents in the spring of 1972 and 1973. For the first time, the Pest Infestation Control Laboratory had evidence that the dressings presented 'an unacceptable risk to wildlife' when used in both spring and autumn (van den Heuvel 1975).

Meanwhile, the Conservancy had used the publicity surrounding these incidents to persuade the Advisory Committee to make a further review. Whilst no one in the Conservancy challenged the need for effective seed-dressings against the wheat bulb fly, the existing situation could no longer be tolerated. Not only was much more research needed to find alternative formulations, but an immediate and complete ban should be imposed on the use of aldrin and dieldrin dressings. This time the Conservancy was even more determined not to allow the Ministry and pesticide industry to 'get away with vague promises, whose real aim is to retain the use of aldrin and dieldrin as seed dressings for as long as possible'.

The Advisory Committee referred the Conservancy's representations to its Wildlife Panel, which concluded in June 1973 that the dressings should be withdrawn unless agriculture could 'produce evidence demonstrating that such action would produce serious economic losses'. It was now the turn of the Scientific sub-committee to proffer advice and, in a paper submitted to the sub-committee, the Conservancy described how Monks Wood had carried out a crash programme on the effects of the dressings since 1969. Attention had been focused on three widely-distributed species, namely the kestrel, barn owl, and sparrowhawk, and on the less common peregrine falcon, short-eared and long-eared owls, and rough-legged buzzard.

The results indicated that almost all the birds found dead or dying, with more than 5 ppm of dieldrin in the liver, came from those parts where the dressings were known to be used in large quantities. Most specimens were retrieved between November and May, and mainly between January and March. None was received between July and

December from outside these areas. The timing clearly reflected both the continued use of the dressings after 31 December and the delayed effects of the legitimate use of the dressings for autumn sowing. Whilst the exact significance of the residue patterns on the population dynamics of the species was unknown, it was very likely that the continued use of organo-chlorine dressings had delayed their recovery following the restrictions imposed on aldrin and dieldrin in 1961 and 1964.[40]

In its consideration of the evidence, the Scientific sub-committee was encouraged to learn that one of the principal suppliers had decided, mainly on economic grounds, to stop manufacturing aldrin and dieldrin dressings from the end of 1973, and had called upon other manufactur-ers to follow suit. This support was forthcoming and, in July 1973, the Advisory Committee agreed with the sub-committee that the withdrawal of all uses of the dressings was most unlikely to 'materially affect national agricultural production. A month later, a press notice from the Minister of Agriculture announced that 'industry has already stopped production of the relevant formulations and supplies to the trade will cease by December 31st this year'.[41]

Once again, the kudos derived from this voluntary agreement was tem-pered by a desire to protect the interests of those farmers who had bought large stocks of the pesticide. In order to give them ample time to use up their stocks, the withdrawal of the dressings was delayed until the end of 1974. Taking the country as a whole, wildlife was unlikely to suf-fer. The Conservancy nevertheless protested at the delay, particularly in view of assurances that only 'small stocks' remained. The Ministry of Agriculture argued that this course was preferable to leaving farmers to dump the dressings about the countryside in an unregulated manner. In the event, the winter sowing season of 1974 was so poor that the dispen-sation was extended for yet a further year. After that time, any remaining stocks were to be disposed of in consultation with experts from the Department of the Environment.

The continuing debate

The threats to wildlife were far from over. The final demise of the orga-nochlorine seed-dressings had been facilitated by the introduction of two new organophosphorus insecticides, chlorfenvinphos and carbophen-othion. Although much more acutely toxic, they were, on balance, safer to use. In November/December 1971, however, over 500 greylag geese (*Anser anser*) were found dead near the confluence of the rivers Isla and Tay in Perthshire. Live birds were seen with foam emerging from the beak; blood was found in their droppings. Biochemical investigations and chemical analyses indicated that the geese had eaten considerable amounts of a mercurial fungicide, of carbophenothion, and of a red dye in proportions that corresponded with those present in a commercial seed-dressing used in the region against wheat bulb fly. Whilst the levels of mercury in the liver were probably too low to have been lethal, ester-ase measurements suggested that the birds had died from organophosphorus poisoning. At that time, there was no published information on the toxicity of carbophenothion to greylag geese available (Bailey *et al.* 1972).

There was no way of preventing the geese from eating the seed; they would always find some corn left on the surface. In the very wet autumn of 1974–75, large numbers of pink-footed geese (*Anser brachrhynchus*) died on Humberside after consuming dressed grain broadcast over the ground, and disc-harrowed. Even germinating grain could present a seri-ous hazard. In the course of investigating three incidents in Angus and Perthshire, field staff from the Department of Agriculture for Scotland found that the geese had uprooted seedlings growing in wet and soft soil, and had thereby ingested carbophenothion (Hamilton and Stanley 1975).

Whatever the merits of replacing dieldrin dressings with carbophen-othion dressings, it was clear that a large proportion of the world's population of greylag and pink-footed geese was endangered. About 65 per cent of the North-west European population of greylag, and 85 per cent of the pink-footed geese, wintered in Britain. The nature conser-vation movement was very encouraged to see that swift steps were taken to prohibit the use of carbophenothion as a winter seed-treatment in Scotland under the Pesticides Safety Precautions Scheme.

In a paper read to a Discussion Meeting of the Royal Society, two members of the Pest Infestation Control Laboratory drew attention to an important difference in the behaviour of organochlorine and organo-phosphorus compounds, and to its bearing on the assessment of environmental hazards under the Pesticides Safety Precautions Scheme. Whereas there was a marked consistency in the toxicity of dieldrin to all animal species tested, chlorfenvinphos and carbophenothion were typical of organophosphorus compounds in displaying a remarkable variation in toxicity between species. It was assumed that the differences reflected the combined effects of comparatively small variations in absorption, meta-bolism, and excretion (Stanley and Bunyan 1979).

It was doubtful whether any pre-clearance tests would have revealed such hazards. Stanley and Bunyan (1979) explained how, in practice, pre-clearance testing had been based on previous experience and was retrospective in design. There were no precedents in the literature, or in field and laboratory studies, to suggest that carbophenothion might be exceptionally dangerous to geese. The incidents had served to highlight

what the Conservancy had been advocating for many years, namely the
need for a post-clearance monitoring system. In their paper, Stanley and
Bunyan described how 'the full implications of the introduction of a
chemical into wide scale use in agriculture or industry cannot be completely evaluated until the chemical comes into unsupervised use'. When
problems arose, there had to be machinery available whereby a multi-dis-
ciplinary study could be made, and the results relayed back to the
registration authority for urgent appraisal and, where required, remedial
action.

The decision of the Royal Commission on Environmental Pollution to
publish a report on *Agriculture and pollution* was an obvious opportunity
to look once again at the whole field of pesticide regulation. Although
the Pesticides Safety Precautions Scheme had been successful in fulfilling
the requirements made of it, the Royal Commission was far from happy
with:

(1) the way pesticides were used on a day-to-day basis, particularly in
    respect of the scale of application;
(2) the quality of the decisions made as to whether pesticides were
    needed, and the type of pesticide to be used;
(3) the adverse environmental effects of excessive use and misuse.

In debating whether the remedy lay in statutory controls, the Commission laid great emphasis on the need to counter the development of
resistance in pest species towards the pesticides being used. The pesticide user could have as much to gain as anyone from the more effective
regulation of pesticide use.

The Royal Commission advocated a two-fold approach. The existing
pattern of control should be left undisturbed for as long as it operated
successfully. At the same time, the Advisory Committee should be put on
a statutory footing, and ministers should be given 'general reserve powers', so that they could introduce mandatory controls over pesticide use
with the minimum of delay whenever changing circumstances required.
This would introduce even greater flexibility into the system. It would
also bring pesticide control in the United Kingdom a little closer into line
with systems employed by some of the member countries of the European Community (Royal Commission 1979).

At a conference in 1980, Professor Norman Moore also argued the
case for retaining a strengthened form of the Pesticides Safety Precautions Scheme. To be effective, any regulatory scheme had to be based on
three elements, namely, flexibility, close surveillance, and the elimination
of abuse. Whatever the faults of the voluntary scheme for clearing individual pesticides, it was doubtful whether another could be found that
was more flexible and based on practicable co-operation. Without the
partial ban on organochlorine pesticides in the early 1960s, he said,

some raptoral species could have become locally, and perhaps nationally, extinct. If the Scheme had been in existence when aldrin, dieldrin, and heptachlor had been introduced, it would probably have prevented their large-scale use, and the story of pesticides and wildlife would have been very different.

Professor Moore believed the main deficiencies of the existing system of regulation arose from its inability to keep the actual use of pesticides under close surveillance. Not only did this make it more difficult to recognize the less obvious effects of pesticides on wildlife, but there were no provisions, for example, for preventing farmers from using excessive quantities of an approved pesticide, despite the damage inflicted on wild-life. There was a strong case for extending the Scheme to include mandatory controls over this kind of abuse, and over the use of strych-nine and other compounds as poison baits. Whilst no laws would ever eliminate such practices entirely, Professor Moore believed that statutory controls could have an important deterrent effect (Moore 1981).

It was not until late 1983 that the Government responded to the recommendations made by the Royal Commission. In doing so, it rejected the need for the general reserve powers outlined by the Royal Commission six years previously. The Pesticides Safety Precautions Scheme had 'long provided very effective practical safeguards', and the powers conferred by the Farm and Garden Chemicals Act of 1967, Con-trol of Pollution Act of 1974, and other legislation, were considered to be adequate. Should circumstances change, and further safeguards be required, these would give 'firmer support' to existing controls, rather than change the way in which they operated (Department of the Environ-ment 1983).

In the last resort, the decision as to how the balance should be struck between voluntary and mandatory controls has to be left to politicians. Britain had demonstrated the value of a voluntary, or self-imposed, sys-tem of controls for dealing with many aspects of pesticide use, but the 'threat' of legislation had always been present in the background. Without access to the highly confidential papers of the pesticide and farming industries, it is impossible to say how seriously the implicit threat of stat-utory controls was taken, but those in the conservation movement believed it was an important factor in winning and retaining the collabor-ation on which the Pesticides Safety Precautions Scheme depended.

PESTICIDES AND POLLUTION

From earliest times, man had taken the presence of particular plant species as an indication of the fertility of the soil. Dogs and geese were used to warn him of the approach of enemies. Now plants and animals were taking on a wider, yet more subtle, role as indicators of change and possibly of danger. In a time of increasing technological change, wildlife could provide an early warning of the problems developing in the environment, which might otherwise remain unperceived until it was too late to remedy them (Moore 1973).

How well equipped were scientists to read the evidence provided by the changing patterns of wildlife distribution and abundance? The studies carried out at Monks Wood on the peregrine falcon and small tortoiseshell butterfly indicated that any simple cause-and-effect relationship would be hard to prove, and unlikely in any case to exist in reality (Moriarty 1972*b*). A wide range of adverse factors impinged on each individual organism. Scientists soon realized that it was impossible to understand the effects of pesticides on wildlife without taking into account other environmental factors, including the presence of other pollutants. Plants and animals were exposed to a 'soup of chemicals'. Murton (1977) described how a watercourse might receive the run-off of farm pesticides, the excess of nitrates and phosphates from agricultural fertilizers, and the contaminants arising from urban and industrial effluent. Few of the constituents of the 'soup of chemicals' were identified, let alone studied in detail. Despite great strides in research, scientists were ill-equipped to advise on the implications of residues found in organisms, whether in the soil, air, or sea. This was graphically illustrated by the Irish Sea seabird wreck in the autumn of 1969.

It is impossible to understand trends in the pesticide/wildlife debate without taking into account the wider concern over pollution. On many occasions, the same organizations and personnel were involved and they drew on the experience gained in pesticide/wildlife research to assess the dangers arising from newly-detected pollutants. An obvious response was to extend wherever possible the various codes of control, originally devised specifically to cover pesticide use. For those still engaged primarily on pesticide/wildlife research, studies of other pollutants were equally informative, providing insights into the passage of compounds through the environment and the behavioural and physiological response of different organisms. They helped to reinforce the impression that the losses inflicted by pesticides on wildlife were no isolated incident. They were rather the harbinger of a new range of factors that would have to be taken into account in the future management of the environment and its natural resources.

This Chapter will look at how an increasingly wide range of actual, or potential, pollutants was causing concern in the late 1960s. It will describe how the presence of polychlorinated biphenyls (PCBs) was first recognized in the environment, and how the compounds came to be implicated in the Irish Sea seabird wreck. First, however, it is necessary to outline how the Nature Conservancy came to be absorbed into a larger research council.

The Natural Environment Research Council

The early 1960s were not only a period when the question of pesticides and pollution began to arouse considerable public disquiet, but also the first time when the future of the Nature Conservancy seemed assured. The two were not entirely coincidental. Until then, the value of having such an official body as the Conservancy had been challenged, particularly by those persons with considerable landed and rural interests. Because the side-effects of pesticides struck first and foremost these same rural interests (whether in the game-covert, fox-earth, or more generally), these critics of the Conservancy were some of the first to benefit from its involvement in the study and dissemination of information on the effects of pesticides.

In this context, a particularly apposite leader appeared in *The Field*, coinciding with the publication of the Conservancy's Annual Report of 1964. The Conservancy's recent achievements had been 'nothing short of immense', and deserved the nation's gratitude.[1] Summarizing the section of the Report on toxic chemicals, the leader commented:

the Nature Conservancy cannot be expected to knock sense into the head of every self-satisfied know-all among farmers who goes his own sweet way regardless of the consequences to those around him, nor to restrain the glib tongue of every salesman paid extra commission on the degree to which he persuades his customers to over-buy. The Conservancy can, however, define the havoc which has been caused beyond all possibility of doubt and challenge, and this it is doing.

There were now misgivings of another sort. Were the terms of reference, and the funds, of the Nature Conservancy adequate to meet the challenges now arising out of the management of the natural environment and natural resources? Was there not a case for subsuming the Conservancy into some larger body, which would be more explicitly equipped to tackle these greater challenges? In 1959, the Minister for Science had asked his Advisory Council on Scientific Policy to review the balance of scientific effort.[2] The Council, which included the Director-General of the Conservancy, concluded that the existing organization of scientific research was broadly satisfactory, except in respect of the conservation of natural resources (Lord President of the Council 1960). Thus, only ten years after the Conservancy had been appointed as a research council, its status was now called into question as part of a much wider review of scientific endeavour.

Right from the start, the Conservancy had been embroiled in a wide range of environmental issues, some of which seemed to be more closely concerned with human welfare than with wildlife. This was a source of strength when it came to negotiating the size of the Annual Estimates with the Treasury.[3] It was through its participation in pesticide research that the Conservancy demonstrated most clearly its potential role in the wider arena of environmental management. By deliberately setting out to discover and highlight the hazards arising from pesticide use, the Conservancy was fulfilling a role which related to, and transcended, its responsibilities for nature conservation.

The fact remained, however, that no organization had a 'clear general responsibility' for resource conservation. Furthermore, the Conservancy's provisional budget of £395 000 for 1959–60 was minute compared with that of the DSIR (£12.8 million), the Agricultural Research Council (5.7 million), and the Medical Research Council (£3.5 million). The Advisory Council on Scientific Policy recommended that a new research council should be created, with a central responsibility for all aspects of resource conservation (Lord President of the Council 1963).

In 1962, the Prime Minister appointed a separate Committee of Enquiry into the arrangements for government-sponsored civil-science (under the chairmanship of Sir Burke Trend). In its report of October 1963, the Committee endorsed the concept of a new Natural Resources Research Council (Prime Minister 1963). A few months later, the functions of the Minister of Education, and of the Minister for Science, were brought together under the newly-created post of Secretary of State for Education and Science. In July 1964, the new Secretary of State announced that a Natural Resources Research Council would be established. The word 'Environment', would, however, be substituted for that of 'Resources' in order to emphasize the wide-ranging terms of reference of the new Council. Under the Science and Technology Act of 1965, the Natural Environment Research Council (NERC) was appointed for the encouragement and support of research in the earth sciences and ecology, the dissemination of knowledge and advice on matters related to these fields, and for 'the establishment, maintenance and management of nature reserves'. The Act transferred the statutory responsibilities of the

Conservancy to the new Council; the Nature Conservancy committee became a 'charter committee' of the Council.

Whilst the interposition of another layer of decision-making meant major changes in the work pattern of senior personnel in the Conservancy, individual scientists perceived little difference in the direction of, and support for their research. As a charter committee, the Nature Conservancy committee still decided its own programme of research. It was difficult for the Toxic Chemicals and Wild Life Section, and for Monks Wood generally, to distinguish the changes brought about by the Conservancy's new status from all the other factors contributing to the uncertainties of the mid-1960s.

The soup of chemicals

The most notorious pollutant in the years leading up to the European Conservation Year was oil. It affected all kinds of wildlife along the coastline, including many species of birds (Parslow 1970). Ever since the early part of the century, the RSPB and other wildlife organizations had campaigned against the evil (Sheail 1976). Although most of the pollution arose from the illegal practice of washing out the ship's tanks at sea, serious pollution could also occur when vessels ran aground, or collided with one another. In 1965, 2 500 tons of oil escaped into the sea as a result of an accident involving an oil tanker in the busy shipping lanes off Beachy Head in the English Channel, but the incident that surpassed all others was the *Torrey Canyon* disaster of March 1967, when 100 000 tons of crude oil escaped into the sea. Ten days after the ship ran onto the Seven Stones rocks off the Isles of Scilly, the British Government ordered it to be bombed from the air in order to burn off the 20 000 tons remaining on board.

Most of the decisions taken during the *Torrey Canyon* incident had a scientific or technical aspect, and the Cabinet Office later published a report compiled by a committee of 15 scientists on these aspects of the disaster (Cabinet Office 1967). The chairman of the committee was the Chief Scientific Adviser to the Government, Sir Solly Zuckerman; the Director of the Nature Conservancy was a member. In the words of their report, 'the disaster was in every respect as unprecedented as it was sudden'. By this time, the Nature Conservancy had become part of the Natural Environment Research Council which, through its various component bodies, also became closely involved in the disaster. Details of the Sites of Special Scientific Interest along the threatened coasts had been hastily compiled by the Conservancy, and circulated to local authorities and other agencies responsible for meeting the oil threat (Fig. 11.1). Inter-tidal plant and animal life suffered as much from the use of

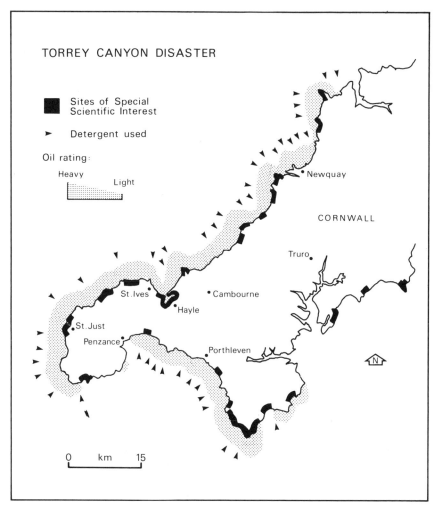

TORREY CANYON DISASTER

■ Sites of Special
   Scientific Interest

► Detergent used

Oil rating:
  Heavy
           Light

Newquay

CORNWALL

Truro

St.Ives
      •Cambourne
    Hayle

•St.Just
Penzance•
          •Porthleven

N

0   km   15

Fig. 11.1 The Cornish Sites of Special Scientific Interest in relation to the incidence of oil pollution and detergent use, following the *Torrey Canyon* disaster.

detergents as from the oil itself (Nature Conservancy 1968). The disaster destroyed the equivalent of the entire breeding population of guillemots (*Uria aalge*) between the Isle of Wight and Cardigan Bay, and a third of the population of razorbills (*Alca torda*) (Cabinet Office 1967).

Despite the interest taken by ornithologists in the general and chronic problems of oil pollution, relatively little scientific investigation had been carried out in the past on how bird populations were affected. Local damage was demonstrably great, but data on the wider and longer-term deleterious effects on the whole population of a species were largely circumstantial and grossly inadequate. In its conclusions, the *Torrey Canyon* Committee recommended not only the further monitoring of populations, but controlled experiments on the ecological effects of oil pollution, detergents, and other methods of removing oil (Cabinet 1967). Experiments to find less toxic dispersants were largely unsuccessful.

Reflecting the resurgence of interest engendered by the disaster, a complete survey of all British seabird colonies was carried out in 1969 and 1970 by the Seabird Group, under a grant from the World Wildlife Fund. The beached-birds surveys, first started in 1921, were reorganized on more comprehensive and systematic lines. To complement these surveys, research was being carried out in various universities on the breeding biology of seabirds and the ecology of auks outside the breeding season, whilst the Conservancy began to study the effects of pollutants, apart from oil, on auks and gulls (Parslow 1970).

Oil was by no means the only ingredient of the 'soup of chemicals' to arouse growing anxiety. Although metal residues occurred naturally in the environment, it seemed likely that they were becoming both more widespread and more concentrated as a result of human activity. There was, however, hardly any information on when and how changes in their distribution and abundance might bring about deleterious effects.

The Toxic Chemicals and Wild Life Section was the first to investigate the environmental effects of lead, when used as an anti-knock agent in cars and other forms of road transport. It seemed likely that small mammals on roadside verges would be most affected by this form of pollution, and analyses were made to determine the lead concentration in the liver and whole body of 101 mammals from three species trapped on roadside verges and in the open countryside in 1969 and 1971. The mean lead concentration in the bodies of the mammals trapped increased from 4.19 d/w ppm on woodland and arable sites, to 5.98 d/w ppm on verges of minor roads, and 7.00 d/w ppm on the verges of the A1, a very busy trunk road. The field vole (*Microtus agrestis*) contained significantly higher total body residues of lead on roadside verges than either the bank vole (*Clethrionomys glareolus*) or longtailed field mouse (*Apodemus sylvaticus*). The difference was attributed to differing food and behaviour (Jefferies and French 1972*b*).

Particular anxiety was felt about mercury, which existed in many forms, both free and bound, with variations in properties and toxicity. Sweden imposed a ban on the use of alkyl mercury fungicides in 1966. The two principal sources of mercury in that country were from the use of phenylmercury acetate both as a fungicide in the wood-pulp industry, and in the chlor-alkali industry. In Britain also, the largest tonnage of mercury was used in industry, agriculture accounting for only 5 per cent. Under the Pesticides Safety Precautions Scheme, curbs had been placed on the use of mercury pesticides, including organo-mercury seed dressings, since 1957.

Reflecting the growing concern over the incidence of heavy metals in the natural environment, and as an extension of the analytical work being carried out at Monks Wood, a pilot survey was conducted to identify the concentrations of mercury in waders and other estuarine species taken along different parts of the British coast. The aim was to determine more precisely the nature of the pollution hazards and to help in identifying suitable avian indicator-species. It was found that, as winter progressed, the concentrations of mercury rose in the livers of the three most numerous wader-species included in the case study of the Wash, namely the knot (*Calidris canutus*), dunlin (*C. alpina*) and redshank (*Tringa totanus*). In the case of the knot, they had risen by 10 to 20 fold by February/March. Parslow (1973) attributed this progressive accumulation of mercury to the nature of the birds' diet, rather than to any seasonal change in the liver physiology or to the redistribution of mercury in the body. It was significant that the levels in another species, the snipe (*Gallinago gallinago*), remained low and relatively stable throughout the year. This species feeds almost entirely in freshwater marshes rather than on the shore. The study implied that the high levels of mercury occurred when the birds were feeding in temperate latitudes and exclusively in estuaries, but were eliminated in summer, when they moved to mainly inland breeding-grounds in the Arctic and Sub-Arctic.

The toxicological and biological significance of the levels of mercury found in the Wash waders was uncertain; there was no direct evidence of any harmful physiological effects. The main source of mercury in those estuaries away from mineralized areas was direct discharge of effluent from manufacturing and refining industries. The Wash, however, received comparatively little trade-waste. Most came from vegetable processing, and presumable contained little mercury. In a later study, Moriarty and French (1977) carried out a limited survey of the sediments of the river and drainage channels flowing into the south-western part of the Wash. They wanted to investigate the significance of the mercurial fungicides used in the bulb-growing industry. Whilst no large scale contamination was found, a sample survey indicated that appreciable amounts of mercury might occur in the ditches around the 200 bulb-dipping sites in the area. Bulb-dipping began at the end of July, and lasted about two months in each year. In the search for an indicator organism that might help in measuring this type of intermittent contamination, the authors of the study found that freshwater molluscs frequently reflected differences in the total mercury present in the sediment.

Many metals play an essential part in the physiology of animals and it is far from easy to determine the extent to which they might, at times, have a harmful biochemical effect. The capacity of animals to take up a foreign metal might also vary seasonally. As in the case of reseach on the effects of organochlorine compounds, the impact of the metals could not be distinguished without a very detailed understanding of the animal's natural physiological cycle.

PCBs

Ever since the first studies had been made on organochlorine pesticides, there had been misgivings as to whether the full spectrum of cause and effect was being investigated. Almost all attention had been focused on aldrin, dieldrin, and heptachlor, and on DDT and BHC, because major losses in birdlife had followed their large-scale use as pesticides. It was an obvious response to use chemical analyses as a way of discovering whether there were residues of these pesticides in the bodies of dead birds found in the field. The analyses appeared to provide the confirmatory evidence, but analysts in the Laboratory of the Government Chemist, and elsewhere, drew attention to the way in which other compounds with a long retention time frequently interfered with the gas-liquid chromotagraphic determination of pp'-TDE and pp'DDT in wildlife specimens (Laboratory of the Government Chemist 1964). Because other forms of halogenated compounds were clearly present, there was the worrying possibility that one or more of these substances might be having a subtle, yet highly significant, effect on wildlife. These suspicions seemed to be substantiated by the discovery of polychlorinated biphenyls (PCBs) as a widespread contaminant (Cairns and Siegmund 1981).

Scientists distinguished the presence and significance of PCBs quite by chance as they began to devise ways of identifying organochlorine insecticides and their persistent metabolites at levels as low as 0.05 ppm in ordinary samples. In doing so, they became even more aware of the peaks corresponding not only to known organochlorine pesticides but to peaks of other unidentified compounds present in the eggs and livers of many seabirds. There could be as many as 10 of these additional peaks on the chromatograph. Roburn (1965) succeeded in establishing that they were organochlorine in nature and, for some time, it was assumed that they represented either a further breakdown or condensation of the organochlorine pesticides, or loose compounds of these substances with such natural products as proteins.

In 1964, Dr Sören Jensen, of the Department of Analytical Chemistry in the University of Stockholm, witnessed up to 14 peaks in chromatograms of extracts from pike, made as part of a study of DDT and its metabolites in human and wildlife fat-tissues. In an attempt to identify these unknown substances, samples of pike were analysed from areas as far apart as Skåne, in the south of Sweden, and unpolluted Lapland. The fact that levels were low in Lapland suggested that the substance did not occur naturally in the fish. Comparative analyses were made of one feather from each of a number of specimens of the white-tailed eagle (*Haliaeetus albicilla*), collected since 1888. The discovery that the unknown substances were already present in specimens from 1942 onwards suggested that they could not be metabolites of the chlorinated pesticides, which only came into use in Sweden after 1945.

Further analyses revealed that the mystery compound was present in increasing quantities as one progressed through the foodchain. The discovery of the corpse of a white-tailed eagle in the islands off Stockholm provided an outstanding opportunity for further investigation. The corpse contained enormous amounts of the unknown compound and, through extensive gas chromatographic studies, the retention times were eventually found to be identical with a group of synthetic compounds, the polychlorinated biphenyls (PCBs) (Jensen 1972).

Not only had no one previously considered PCBs as contaminants in the environment, but there seemed every possibility that some of their effects on wildlife might have been attributed mistakenly to DDT and other pesticide residues. It was also clear that any attempt to regulate the use of PCBs would present an entirely different range of problems from those encountered in the use of pesticides. Whereas pesticides had been used in the open environment with intent, PCBs had entered by the 'back door'. There were at least four possible routeways: via the chimney stacks of manufacturing plants; as consituents of industrial effluent; through the gradual weathering or wear of products containing PCBs, or as a result of the incineration of discarded products containing PCBs.

PCBs were first used in America in 1929 and introduced into Europe about 10 year later, mainly as protective coatings, plasticizers, sealers in water-proofing, printing inks, and synthetic adhesives. In liquid form, they were used as hydraulic fluids, and in thermostats, cutting oils, and grinding fluids. They were also incorporated in electrical apparatus. Those qualities which made them so valuable for industrial use, namely their chemical stability and insolubility in water, also made them potentially dangerous as environmental pollutants.

News of the discovery of PCBs in Baltic marine life appeared first in the *New Scientist*, as a short note published under the heading, 'Report of a new chemical hazard' (Anon. 1966). In Britain, a paper by members of the Laboratory of the Government Chemist confirmed that the long-retention compounds previously recognized were indeed those of PCBs (Holmes, Simmons, and Tatton 1967). Although no one realized it at the time, the ability of PCBs to escape into the environment had already been demonstrated in the laboratory at Monks Wood. Moriarty (1975b) recalled later how he had designed a controlled experiment to test the biological effects of dieldrin on some caterpillars. When cleaned-up extracts from the caterpillars were run through a gas-liquid chromatograph, a smooth base-line was expected from the control population of 'undosed' caterpillars. There were instead so many large peaks as to mask entirely the peak for dieldrin from the dosed caterpillars. The experiment had to be abandoned; no one could explain what these peaks represented, or how the caterpillars had come by the compounds which had caused them. It was only later, when the analyst at Monks Wood, Mr M.C. French, saw some of the first published chromatograms for PCBs that he recognized the pattern. Decorators had been re-varnishing some of the exterior woodwork of the laboratory, and presumably the caterpillars had picked up some PCBs from the varnishes, despite their being in a room with no open windows and a door that opened into the middle of the building.

There was soon plenty of published evidence for the widespread presence of PCBs in the environment. Jensen, Johnels, Olssen and Otterlind (1969) described, in a paper in *Nature,* how, on the basis of 176 samples taken from the Swedish marine ecosystem, they had found the Baltic to be significantly contaminated with PCBs—their presence had not even been suspected three years previously. American scientists found that PCBs were as widely dispersed as organochlorine pesticides in the transatlantic environment. Their precise distribution and concentration in marine and terrestrial ecosystems depended on prevailing patterns of wind circulation and rates of fall-out (Risebrough, Hugget, Griffin and Goldberg 1968). In a further paper in *Nature* , Risebrough, Rieche, Herman, Peakall, and Kirven (1968) warned that together with other chlorinated biocides, PCBs could account for a large part of the aberration in calcium metabolism witnessed in many species of bird since the war.

Meanwhile, steps were being taken at Monks Wood to ascertain the extent and significance of contamination by PCBs in British wildlife. In a paper published in the first issue of the journal, *Environmental Pollution* (edited by Kenneth Mellanby), Prestt, Jefferies, and Moore (1970) described the results of analyses made on 196 livers from 33 species, and 363 eggs from 28 species of bird, obtained between 1966 and 1968. Most of the species were predatory. PCBs were found in terrestrial species from most regions, and in most eggs from freshwater species collected from the Midlands, and east and south of England. The residues were also present in all the individual and bulked samples of seabird eggs examined from a west-coast and two east-coast colonies. The highest liver residues were found in freshwater fish-feeding birds and bird-feeding raptors. A level of about 900 ppm of PCBs in a heron from

Derbyshire, and 300 ppm in another from Settle in Yorkshire, proved to be the highest recorded residues for PCBs in any part of the world. The level was similar to that of pp'-DDE.

In view of these findings, it was clearly relevant to discover whether PCBs could have been responsible for the effects previously attributed to organochlorine pesticides. In order to assess avian toxicity, samples were secured of the compounds being produced commercially in Britain, and fed to Bengalese finches. It was estimated that a dose rate of 254 mg/kg/day was required to produce 50 per cent mortality at 56 days. The mean liver content was 345 ppm. On this basis, the toxicity of PCBs was calculated to be only one thirteenth of that of DDT, which was, in turn, only one sixteenth to one fourteenth of the lethal chronic toxicity of dieldrin. Prestt *et al.* (1970) concluded that very few of the specimens examined during the course of their surveys were likely to have died from either PCBs (or DDE) poisoning.

There was still every indication that aldrin, dieldrin, and heptachlor had caused the sudden decline in certain birds of prey. This had coincided in space and time with the introduction of these pesticides, whereas PCBs had first been used before the Second World War. Although wildlife had been increasingly exposed to PCBs, there was no indication of any sudden increase in exposure during the late 1950s. There were, however, grounds for asserting that PCBs had an important sub-lethal effect. One of the symptoms of PCBs-poisoning was hydropericardia, namely the presence of a fluid in the pericardial sac causing a back pressure on the heart. This symptom was noticed in both the dosed Bengalese finches (Prestt *et al*, 1970) and in the guillemots later found dead in the Irish Sea (Holdgate 1971). In the laboratory, those birds dying from PCB-poisoning had enlarged kidneys, and some of them, before death, displayed apparent leg paralysis, or body and wing trembling. These trials, and those conducted elsewhere on pelicans and cormorants, suggested that PCBs might be one of the causes of breeding failure (Anderson, Hickey, Risebrough, Hughes, and Christensen 1969). Such an effect could impede any recovery in the status of birds of prey, following the reduced use of organochlorine pesticides.

The incidence of PCBs in marine life around the Scottish shores was studied by the Freshwater Fisheries Laboratory at Pitlochry. The highest concentrations were recorded in the Firth of Clyde. In an attempt to establish how these residues reached the fish, attention was focused on the crude sewage sludge from Glasgow and adjacent areas, which was transported by boats to the estuary and dumped in deep water south of the Isle of Bute. An average concentration of 1 ppm was found, equivalent to a discharge of 1 ton of PCB isomers per year in the Clyde. A similar level of discharge was reported in the northern Irish Sea, arising from the dumping of crude sludge from Manchester. Because of the manifold uses of PCBs, and the lack of data on how they were exploited in the several hundred factories served by the sewers, it was impossible to identify the industrial sources of the 'contaminated' sludge (Holden 1970).

It was apparent that wherever sludge disposal occurred, whether at sea or on land, there was a possibility of PCBs being released into the environment. In America, Risebrough (1970) outlined the conservationist's concern in an article published in *Environment*. As an addendum to the article, the journal published a statement made by the sole manufacturer of PCBs in the United States, Monsanto Chemicals. The statement expressed concern over the findings of Jensen and others, but stressed that PCBs, like many other industrial chemicals and home products, were not hazardous when handled and used properly. The precise source of the residues identified as PCBs in marine life was still not known, and it would 'take extensive research on a world-wide basis, to confirm or deny' the scientific conclusions that had been reached and published. The company was co-operating fully in these studies.

Following correspondence and an exchange of information in the early part of 1970, a meeting was arranged at Monks Wood, which was attended by members of the Toxic Chemicals and Wild Life Section and a representative of NERC, and by personnel from St Louis, Brussels and north Wales, representing Monsanto chemicals.[4] The Conservancy and NERC had already been struck by the helpfulness of the company, but no one looked forward to the meeting. It was expected to follow the well-established course of those relating to pesticides, where representatives of industry challenged analyses, and denied the significance of any level of residue proven beyond doubt. Those in the Conservancy knew only too well that no one could prove that PCBs harmed wildlife—there was just the worry that they might do so. In the event, the meeting was so harmonious that, paradoxically, only a single copy of a handwritten record survives in the Conservancy's files of what must have been one of the most significant meetings ever to take place at Monks Wood.[5] It was an example of harmony generating far less paperwork than discord.

John Parslow's note on the meeting indicated that the company's representative, in charge of co-ordinating studies of PCBs in the environment, opened the discussion. He told how pollution was a highly topical issue in the United States and Canada, and how Congressman Ryan had called a press conference and issued statements saying that PCBs were both dangerous and widespread. Like most people, Ryan's main concern was about the possible effects on man. The representative of the central medical department at St Louis recalled how the company had first become concerned in the autumn of 1968, and how, by the following spring, it had accepted evidence for the presence of compounds in wildlife specimens as irrefutable. Monsanto Chemicals had sought, as a matter of urgency, information on the biodegradability and toxicity of the compounds. Having reviewed the most recent analytical data available, the meeting at Monks Wood agreed that there was cause for concern, particularly in view of recent experience with pesticide residues. The company's representatives emphasized that, unless industry introduced its own controls voluntarily, there was every possibility of governments imposing those controls. PCBs should only be used in 'controlled' environments. Fortunately, over two-thirds of the PCBs manufactured in the United States were used in the electrical industry where control was possible, but, even here, it was essential to prevent spent fluids, or those from damaged and discarded equipment, being tipped down the sewers or dumped at sea. The aim should be to find substitutes for PCBs wherever possible.

First in America and then in Europe, Monsanto Chemicals announced that it would stop supplying PCBs for 'non-controllable' uses, such as the manufacture of lacquers, paints, lubricants, and paper. The only permissible uses would be in closed systems, where the material would remain confined, namely in dielectric fluids in transformers and capacitors. There were no adequate substitutes for these uses. Because the company would take back the spent fluids, there should be no danger of any of the material escaping into the environment.

The Nature Conservancy was further encouraged when it learnt of the correspondence passing between Monsanto Chemicals and its customers, and within those customer organizations. In correspondence dated as early as February 1970, the company drew attention to reports in several newspapers and magazines that PCBs had been found 'in some marine, aquatic and wildlife environments'. The letter identified which of the company's products contained the compounds, and set out ways of minimizing the kind of risks identified. One of Britain's largest manufacturing industries issued an internal memorandum to those handling electrical apparatus, describing how the compounds, even in small quantities, could 'have disastrous effects on certain wildlife environments'. It continued:

they can be handed on from prey to predator where they lodge in body tissue. Being almost indestructable the dose accumulates as more contaminated prey is eaten until a lethal level is reached. There is ample evidence to show that a closely allied group of substances used in pesticides had in recent years and by the same process gone close to wiping out by sterilization or ultimate death many of the species of birds of prey in Britain and elsewhere. Marine contamination has been detected and it is of growing concern as to what this might lead to unless it is checked at once. So much human food has a marine origin.

The voluntary withdrawal of specified applications of PCBs was widely welcomed by the conservation movement, not only as a contribution towards environmental health but as a precedent for public spiritedness on the part of a multi-national company with a virtual monopoly in the supply of the product. Fortunately, fears that another manufacturer might seek to replace Monsanto Chemicals as a supplier of PCBs were not realized. In view of the large amounts of the contaminant already in the environment, many years were likely to pass before seabirds and their eggs were entirely free from the residues.[6]

The Irish Sea seabird wreck

Once again, an incident involving heavy loss of birdlife rivetted public attention on the actual and potential effects of a pollutant and provided further circumstantial evidence of the insidious threat to wildlife posed by chemicals 'released' into the natural environment. In view of the frequent forebodings uttered by the Conservancy and the conservation movement, the Irish Sea seabird wreck in the autumn of 1969 provided a salutory insight into their capacity to respond when an incident occurred. There were important lessons to be learned in both the adminstrative and research fields.

The first lesson was that not all incidents came with the suddenness and drama of the *Torrey Canyon* disaster, when the dangers and their causes were immediately apparent. At first, there was nothing to indicate that there was anything exceptional in the scale of mortality or its geographical extent in the Irish Sea. The calendar of events began in late August, when a sick seabird was found in Strangford Loch. It died two days later. Over the following three weeks, the easterly winds brought many more dead and dying birds ashore. It was not until 23 September that news of these losses was relayed to the headquarters of the RSPB.[7]

A single sick guillemot was seen in the Menai Straits on 20 September. When further dead and weakened guillemots were swept onto the beaches of west Anglesey on 25 September, it was assumed that they had been in contact with two oil slicks sighted off South Stack a few days earlier. It was not long, however, before it was noticed that many of the birds were free of oil, in full moult and emaciated. The Conservancy's headquarters learned of their puzzling condition on 1 October. Some carcasses had been sent to the MAFF Veterinary Laboratories at Lasswade but no evidence of disease could be found. On 9 October, the regional office for south-west Scotland learned from the Seabird Group that 1 500 guillemots had been discovered, washed up on the beaches of Ayrshire.[8]

The lack of 'people on the ground' made it difficult to keep the coast of south-west Scotland under surveillance. Neither the RSPB nor the Conservancy had any outpost staff, and the absence abroad of the Assistant Regional Officer based at Loch Lomond, meant that the Conservancy's nearest personnel were in Edinburgh. Because the holiday season was over, most of the seabird corpses passed unnoticed; those found were quickly buried by local authority workers. The first casualties came ashore in small numbers in the last week of September, Then following the severe south-westerly gales of 27–28 September, the number rose to a maximum in the week beginning 13 October.

The voluntary organizations were the first to suspect that the whole of the Irish Sea had been affected by a seabird wreck. Following a telephone call from the Director of the RSPB on 10 October, Mr W.D. Park of the Conservancy's headquarters called for detailed reports from the Conservancy's various regional offices. Only then did the real significance of the locally reported losses become clear. Fortunately, the peak of mortality seemed to have passed by that stage.

Although the voluntary bodies were keen to issue a press statement, it was decided to wait until 'a more rounded story' had emerged. It was not until 15 October that the RSPB issued a press release, describing how at least 8 000 birds had been found dead and dying, mainly on the Ayrshire coast. Although most had probably died in the gales of 27–28 September, when winds of up to 96 miles per hour were recorded at Prestwick, the unprecedented scale of the wreck was puzzling. Autumn gales were by no means unusual, and it was the larger, rather than the smaller, auks that were most affected. In most parts, over 90 per cent of the casualties were guillemots. Most were emaciated adults in wing moult. The press release, and a circular letter addressed to members of the RSPB, appealed for more information.[9]

It was not long before the national press began to make a drama out of a crisis. Under the headlines, 'War dump may be bird killer', *The Times* of 18 October recalled how sealed canisters and shells, dumped in deep water in the Baltic Sea after the war, had recently become a dangerous hazard. Things were not improved on Monday, 20 October, when the *Daily Telegraph* published an article, suggesting that 'insecticides may have killed birds'. Following a consultation at the Conservancy's headquarters, the Deputy Director (Research), Dr M.W. Holdgate, appraised the headquarters of NERC of developments. As the Secretary of the Council, Mr R.J.H. Beverton, later recalled:

I personally decided that the situation called for a general alert, even though at that stage the results of the analyses were not yet available, and there was no evidence that the incident was due to other than natural causes. On the same day, I informed the Directors of all the N.E.R.C. marine laboratories of the urgency: I instructed them to check their records for marine fauna generally, and to put in hand at once such additional surveys as were material and feasible. Within hours one of these establishments reported that there was nothing unusual about the plankton of the area.[10]

On the following day, the 'leading ornithologist', James Fisher, was quoted as saying that 'this is the biggest seabird crisis of the present generation, if not ever'. The RSPB warned that the extinction of certain breeding colonies was likely. Some bird-experts believed the mystery killer might be a world-wide virus epidemic. There were reports of dead seabirds and grey seals being sighted on the Cornish coast. Dr Moore of the Monks Wood Experimental Station was quoted by the *Daily Telegraph* as suggesting there might be a range of causes. No one could be certain until further carcasses had been examined.

Apparent disarray among naturalists and scientists as to the cause of the deaths, and the time taken to collect and analyse specimens, gave support to criticisms voiced by Anthony Tucker in the *Guardian*. Under the headline, 'Dead birds, dead loss', Tucker wrote:

The scale of the kill was not recognized early enough, and the scientific backing is too patchy for any conclusive outcome to be expected. With environmental emergencies occuring ever more frequently, is it not time we looked seriously into the question of setting up a properly backed national emergency service? We have the laboratories, but they are not being used.

The article marked an important shift in press coverage of the wreck. From concern over the wreck itself, press attention was now turning to the alleged inadequacies of the Government's response to that wreck. In the eyes of the press, NERC and the Conservancy were part of the government machine.

The situation had become so urgent and so serious that Mr Beverton gave 48 hours' notice of a meeting at NERC headquarters in order to review the evidence which was now being gathered more quickly. Representatives of the 'official side' and of the voluntary bodies would meet under the chairmanship of Dr Holdgate. The draft agenda identified three principal items for discussion, namely:

(1) a critical review and evaluation of existing data;
(2) the need for additional data and action;
(3) the extent to which new mechanisms for concerted action in future incidents of this type might be devised.

Unknown to anyone at the Conservancy's headquarters, a new item was to appear on the agenda by the following morning.

A swirl of dirty water

The first sample of eight guillemots from Ayrshire had arrived at Monks Wood, and Dr Jefferies, Michael French and John Parslow began to carry out autopsies and analyses straightaway. On the evening of 21 October, Michael French showed them 'the perfect fingerprint' on the gas chromatograph of one of the commercially-produced PCBs that had been used in the earlier feeding trials with Bengalese Finches. A range of 130 to 450 ppm of PCBs was recorded in the livers of the eight birds. The normal lethal dose recorded in monitoring and experimental work was 250 ppm. Next morning, the three of them saw Dr Mellanby, and suggested that 'a quick paper' might be published. They realized that the findings might be a little 'hot', and everyone agreed that Dr Mellanby should telephone the Conservancy's headquarters in London. All stood round in Dr Mellanby's office, 'wondering what they had stirred up', as Dr Holdgate was informed of their findings.

After this telephone call, Dr Holdgate had urgent consultations with the Chairman of NERC, Professor V.C. Wynne-Edwards, and Mr Beverton. All expressed concern over the small size of the sample. If the birds had died from ingesting PCBs, these must have arisen from some local incident. It seemed extremely unlikely that the very large number of casualties could be attributed to the same source. As Professor Wynne-Edwards commented, 'the rate of dilution of PCBs was likely to be so great as to make it virtually incredible that any single pollution incident could cause mortality among seabirds over the whole distance from Oban to Cornwall without there being other obvious alarms'. It was agreed 'that it would be most unfortunate to give publicity at the present stage to these preliminary results'. Every effort had to be made to carry out further determinations, and to include birds of other species from other localities. Monks Wood was instructed to treat the results of the analyses as an official secret.[11]

Urgent steps were taken to rush further carcasses to Monks Wood and other laboratories by car. The Laboratory of the Government Chemist and the Infestation Control Laboratory were told in confidence of the analyses made at Monks Wood. The Fisheries Laboratories at Lowestoft and Aberdeen were asked to check recent landings of fish in the affected areas for signs of contamination. In the course of these telephone conversations, Dr Holdgate learned of concern expressed over PCBs at the recent International Committee on the Exploitation of the Sea.

In the light of these developments, the meeting planned for 24 October assumed even greater significance. In the words of a NERC press notice, it was attended by representatives of the voluntary and statutory bodies concerned with birds, the NERC research institutes, the Fisheries Laboratories, the Government Chemist, and other interested government departments. It established that abnormal numbers of dead birds had been found on both sides of the Irish Sea. There was no confirmed evidence of seals, fish, or other marine life being affected.

It was considered that many factors might have contributed to the incident. At that time of year, the auks were moulting and were vulnerable to extra stress. Severe gales had occurred during the peak of mortality, and had undoubtedly driven many birds ashore. About a fifth were oiled, some heavily. Although some birds showed signs of internal lesions and respiratory infection, pathological examinations had revealed nothing particularly abnormal. A paragraph in the press statement described how analyses had been carried out on eight guillemots, and a further bird captured alive in Northern Ireland. Most had higher levels of PCBs than had been detected in earlier, routine monitoring. Whilst the toxicity of PCBs to seabirds was unknown, experience with other bird species suggested that these higher levels might be significant.

The main value of the meeting was the elimination of false lines of enquiry, and the focussing of attention on potential sources of infection or pollution. A columnist in *Nature* warned of the risk of it becoming fashionable to make a fuss whenever PCBs were found, irrespective of the fact that practically any industrial compound could be discovered in the environment if sought by the sensitive measuring devices now available. No direct links had so far been established between PCBs and the Irish Sea wreck. The lack of any direct evidence was also stressed at a meeting of the Council of NERC on 28 October.[12] The only significant indication of some kind of correlation between population decline and PCBs arose from analyses of pesticides and PCBs found in a variety of marine organisms in Swedish coastal waters, during which 'exceptionally large amounts of residues' were found in the white-tailed eagle (*Haliaeetus albicialla*), a species which had suffered a severe decline in numbers over the previous few years (see p. 207). The levels recorded were similar to those present in the sample of guillemots from the Irish Sea.

The meeting invited Dr Holdgate to co-ordinate a further investigation, involving at least six laboratories. Although there was no positive evidence of any significant disturbance in the marine ecosystem, samples of fish, plankton, and sea water should be taken. A follow-up meeting was convened by NERC on 11 November 1969. A further press notice identified the 27 voluntary and research bodies represented, and the seven government departments, including the Cabinet Office, which sent observers.

This meeting confirmed the scale and character of mortality, and heard that there was no concrete evidence of any viral or bacterial infection in the 50 or more carcasses examined. Analyses of 36 birds for heavy metals and organochlorine residues had revealed nothing of significance. The levels of PCBs varied widely, but were relatively high in relation to pesticide residues when compared with the ratios found in other wild birds. The press notice stressed that 'insufficient is known about the distribution, physiological action and toxicity of these substances to

Guillemots to draw any firm conclusions as to their significance as a contributory cause of death'. The meeting again concluded that 'no single
factor, natural or artificial, stands out as being the probable cause of this
incident'. A high priority would be given to intitiating further, co-ordinated studies.[13]

The statement did nothing to allay mounting press criticism of the
response to the seabird wreck. In the *Observer* of 26 October, the leader
writer had recounted how:

thousands of dead sea birds were washed on to the shores of Britain. A week
later, someone realised there had been a major environmental disaster. After a
further week, one bird had been analysed, but only for a few chemical pollutants.
After yet a further week, a dozen birds had been studied to test a few other theories, and the Government, at last, called some experts together. Meanwhile, any
further evidence they might need to solve this mystery had been rotting on the
beaches.

So far investigations had depended on the efforts of the voluntary bodies
and whatever friendly contacts they happened to have with a few laboratories. A national environmental research unit was required, presumably
under the new Secretary of State for Local Government and Planning,
which could 'collect the evidence quickly, co-ordinate the detective work,
and have access to suitable laboratories as of right'.

Such criticisms took an acutely embarrassing turn for the worse when
the *Observer* of 9 November 1969 quoted a statement by the RSPB
Press Officer on the 'mass pollution problem in the Irish Sea'. The statement concluded with the words:

The Nature Conservancy says it has nothing to do with them, because they do not
go down to the sea . . . What investigation is being undertaken is fragmented and
is not properly co-ordinated or directed by one central authority.

NERC and the Conservancy were appalled that a voluntary body should
wish to criticize the Conservancy so publicly, particularly when the allegations were patently so untrue. The RSPB was equally horrified by the
article; its Press Officer had never made the statement. A letter pointing
this out was despatched immediately to the *Observer* and was printed on
the following Sunday, together with an apology from the newspaper,
admitting that there had been an error in sub-editing. The article had
been composed in Manchester, and had been carried in full by the northern editions of the paper. A paragraph had been omitted from the
London editions, which would have made it clear to readers that the
highly critical comments had been uttered by a marine biologist, and not
the RSPB. Not only did the blunder remain uncorrected for a week, but
worse was to follow. *The Times* went on to the offensive two days after the
meeting of 11 November. Not only had there been a failure to mount an

immediate investigation, but there was still no official explanation of how the abnormal concentrations of PCBs came to be in the Irish Sea.[14]

Although most of the criticism was not directed specifically at NERC or the Conservancy, an explanation of their course of action was required. Right from the start, the Department of Education and Science had emphasized that the Research Councils should look only for the cause, and not for the source, of the pollution. They were not policemen. It was the responsibility of other Government Departments to eliminate pollution and, throughout the period of the wreck, the Conservancy simply passed any information on the dumping of effluent in the Irish Sea to these departments.[15]

In a long letter to the Office of the Secretary of State for Local Government and Regional Planning, Mr Beverton set out the responsibilities of NERC in the seabird incident. The principal role of the Nature Conservancy within NERC, he wrote, had been to provide an official point of contact and such facilities as those of the Toxic Chemicals and Wild Life Section at Monks Wood. It would have added enormously to the demands on manpower and expenditure if the efforts of the many hundreds of amateur ornithologists up and down the country had been duplicated. The voluntary bodies were, and should remain, responsible for much of the field survey and research work on birds.[16]

Mr Beverton added that he doubted whether the evidence available could have justified a more rapid response. It was important to maintain a sense of proportion if 'we are not to be pressurised into precipitous action verging on the hysterical, with the possibility of missing the really important happenings'. If a nation-wide exercise were to be mounted every time anything untoward was noticed in the natural environment, 'we would be wasting enormous resources of money and skilled scientific man-power chasing imaginary scares'. It was still 'the judgement of the best informed scientists' that the case for PCBs being the main cause of the Irish Sea seabird wreck was far from proven. NERC intended to publish a full report on the incident and to extend monitoring and research on certain of the more significant residues in seabirds, fish, and other marine organisms.[17]

It was another month before the press finally lost interest in the seabird wreck. On 18 November, Anthony Tucker described in the *Guardian*, under the headline, 'A swirl of dirty water', how Britain in the 'Age of Effluent' was lagging behind other countries in the effective control of pollution. Somewhat belatedly, the *Daily Mirror* devoted its front page on 8 December to the 'Riddle of "the Ditch of Death"'. The need for a definitive account of the incident and its investigation remained just as urgent as ever for NERC. Not only had the Secretary of State for Local Government and Regional Planning called for such a report, but the appointment of a Royal Commission on Environmental Pollution had been announced (see page 2). Particularly in these circumstances, it was vital that NERC asserted its role in the pollution field. In the words of the Council's Secretary, 'to fail on this exercise would be disastrous'.[18]

NERC was to edit and publish the report. The factual sections would be based on the reports submitted by the 30 departments, research institutions, and voluntary bodies that had taken part in the incident. In the final section, Dr Holdgate would evaluate and interpret the evidence, and indicate where any of the contributors might dissent from the conclusions reached. As Mr Beverton informed Monsanto Chemicals in February 1970:

our aim in this report is to set out the facts and their interpretation strictly impartially, endeavouring to assign proper weight to what we know and what we do not know. I hope in so doing we shall be making a modest contribution to the preservation of a reasonable sense of proportion in what is a highly emotive subject.[19]

Deadlines for the submission of drafts came and went. The delays stemmed from many factors, not the least of which were the competing demands on the time and resources of the scientists involved. During the period, October–December 1969, a total of 240 man-days had already been taken up with the survey, collection, analysis, research, and administration in connection with the wreck. Research programmes at Monks Wood and Merlewood had been disrupted, and routine work in five of the Conservancy's regions affected. For many investigators, it was a frustrating assignment. So little 'hard information' was available on past seabird wrecks, and on wind and water movements in the Irish Sea. The impossibility of obtaining fresh, but affected corpses, once the peak of the wreck had passed, precluded the pathologists of the Central Veterinary Laboratory and elsewhere from providing more definitive information.[20]

The difficulties involved in mounting a sampling programme to cover marine organisms were set out by Dr B.B. Parrish. Even if PCBs were involved, he wrote, no one knew where to look for their principal source. Had a cargo vessel, carrying PCBs either up or down the Irish Sea and the Clyde, leaked material? Had a major spill, intentional or otherwise, occurred in one locality, and the birds become contaminated and subsequently dispersed as a result of the gales and normal migration? Had there been a gradual accumulation of PCBs in food, presumably as a result of some kind of industrial discharge? In that case, had the build-up 'led to sickness and mortality over a fairly wide area', which had then become markedly apparent as birds were driven ashore by the gales? Each hypothesis called for a different sampling programme.[21]

In its draft submission of June 1970, the Toxic Chemicals and Wild Life Section continued to argue that starvation due to weather conditions might have caused a loss of body weight, reducing the amount of fat and thereby increasing the amount of circulatory PCBs. The Central Veterinary Laboratory, and Department of Animal Pathology at Cambridge, confirmed that the amount of PCBs in the liver were high enough to cause lesions similar to those found at Monks Wood in dosed birds during earlier experimental programmes.

Considerable thought had to be given to the presentation of the report. The conclusions were so tentative and negative that they might easily give the impression of the report being 'Much ado about nothing'. NERC decided to publish a glossy, short report, giving the essential facts and main conclusions, together with a longer report, containing the analytical data in mimeograph form (Holdgate 1971).[22]

Those readers looking for a neat and convincing solution to the mystery of the wreck were certainly disappointed. The report stated that, because of the wide distribution of dead birds, it was unlikely that all the victims had died from a single cause operating in one place at one time. Of the range of primary causes considered, disease or a shortage of food seemed the most likely. The main effect of the winds may have been to increase existing stress and to drive the corpses on to the beaches. Furthermore, any interpretation of the post-mortem findings was made difficult by the small number of specimens collected in relation to the estimated 100 000 birds that might have died. Considerable significance was, however, attached to the histological examinations of the Central Veterinary Laboratory, which revealed a consistent pattern of unusual changes in the liver, resembling the lesions observed in chicks fed experimentally on PCBs.

The NERC report recalled how Monks Wood had found unusually high concentrations of PCBs in the first carcasses analysed from the wreck. These levels were higher than those of organochlorine insecticides—ratios unlike those previously recorded in land and freshwater birds. There was, however, no evidence of unusually high concentrations in other marine life in the Irish Sea, and the *total* amount of PCBs, DDE, and dieldrin in the bodies of a small sample of healthy birds shot outside the area was at least as high, if not greater, that that recorded in the wreck. The most striking difference was that the residues in victims of the wreck were highly concentrated in the birds' livers. In view of the emaciated condition of many of the birds, it was reasonable to suggest that PCBs and DDE were mobilized as a result of the decline in body- or liver- fat reserves, and that they had passed to the liver and other organs, where analogies with Bengalese finches, chickens, and quails fed PCBs under experimental conditions indicated that the residues could cause the type of damage revealed during post-mortem examinations.

The NERC report concluded that the presence of PCBs could have added to the stress imposed by storms, moulting, and starvation, and 'have loaded the odds against recovery'. The pollutant did not, however, explain the initial onset of malnutrition or interupted feeding, particularly in view of the apparently healthy state of birds shot in other areas with equally heavy total-body loads (Holdgate 1971).

Whilst the mystery of the seabird wreck was never fully resolved, subsequent research at Monks Wood added weight to the possibility of PCBs being more directly implicated. During the drafting of the report, the experimental evidence available suggested that the livers of birds fed PCBs contained only a very small percentage of the total-body weight of PCBs, and that its correlation with dose rates was poor (Prestt *et al.* 1970; Parslow, Jefferies and French 1972). In a further series of analyses, Parslow and Jefferies (1973) examined the relationship between liver and total-body residues in two groups of guillemots, namely one group of eight adults found dead and in wing moult during the seabird wreck, and another group of birds shot in November 1969 in the north-western part of the Irish Sea. In this experiment, a significant relationship was found between liver and total-body residues. Using this knowledge, the authors of the study then calculated, from known amounts in the livers, the total-body loads in a larger series of 39 guillemots also found dead at the time of the wreck. These birds had nearly twice the mean level of PCB residues as those shot in an apparently healthy condition. These findings led the Toxic Chemicals and Wild Life Section to be even more convinced that PCBs had played a significant role, together with such factors as food shortage, in the seabird wreck.

Monitoring the environment

There was no doubt that the press and other critics would expect a much more immediate and effective response when the next wildlife disaster struck. As early as October 1969, discussions were under way to see how incidents might be recognized and assessed more quickly. Urgent consideration was once again given to how 'base-line environmental data' might be collected as part of a comprehensive and detailed monitoring programme, and how that information might lead to more effective action.

By the mid-1960s, the term 'monitoring' had come into general use, particularly in connection with the recording of radio-active fall-out from nuclear tests (Moore 1975). When the President's Science Advisory Committee in the United States recommended the extension of the monitoring of pesticide residues in its report of 1963 (see p. 96), the Nature

Conservancy advocated the introduction of such a system in Britain, designed and organized by the Conservancy.[23]

In its submissions to the Advisory Committee and to the (Frazer) Research Committee, the Conservancy had emphasized that a monitoring scheme could help in assessing the benefits of the partial ban imposed on the use of organochlorine pesticides in 1964. It was now possible to measure with considerable precision the presence of residues which, only a short time previously, would have passed undetected. The Laboratory of the Government Chemist and the Department of Agriculture for Scotland had developed a simple and rapid method of extracting the major organochlorine residues, and of determining their level by gas-liquid chromatography. The Toxic Chemicals and Wild Life Section had used a semi-quantitive paper-chromatographic method since 1962 as a confirmation technique. As early as October 1963, there had been discussions on ways of improving collaboration between laboratories in order to make it easier to compare the results obtained (Laboratory of the Government Chemist 1974).

Whilst a great deal of worthwile data had been obtained, and the gross quantitive differences in residue levels could be compared, the research base was still extremely imperfect. The (Frazer) Research Committee drew attention to the lack of suitably trained biochemists, pharmacologists, toxicologists, and animal pathologists. The instrumentation needed to keep abreast of advances in analytical chemistry was extremely expensive Furthermore, the sample of the wildlife population was often biased; the Toxic Chemicals and Wild Life Section had to depend on local naturalists finding dead or dying birds in the field, so that examination and analysis had to take account of the likelihood of post-mortem changes. It was known that pp'-DDT underwent reductive dechlorination to pp'-TDE (Walker, Hamilton and Harrison 1967).

In view of these and other material constraints, the (Frazer) Research Committee believed that the setting up of a national monitoring programme should be postponed until the results of on-going surveys had been received, and the toxicological significance of any residues assessed. Moore (1966b) conceded that a monitoring system designed to measure changes on a world-wide basis was probably premature on scientific grounds, let alone administratively, but the impediments should not prevent everyone from working towards that goal.

The appointment of the Royal Commission on Environmental Pollution, and of the Central Unit on Environmental Pollution in 1970, provided further opportunities to review the steps being taken to monitor changes taking place in the environment. An extremely complex situation was revealed. Over a hundred programmes covering land, water, and air pollution were identified. Although more was probably known about the

British environment than about that of any other large industrialized country, programmes were still designed with little regard to the way in which they might contribute to the *overall* picture of pollution. There was need 'to strengthen the machinery for co-ordinating the programmes in different sectors' so as to ensure that the data were collected in a compatible form and to meet 'domestic policy requirements' (Department of the Environment 1974*a*).

At a time when the Government's interest in environmental issues was increasing, there was no shortage of suggestions as to how improvements might be made in monitoring and managing changes in the environment. The Irish Sea seabird wreck had demonstrated extremely well how such incidents could impinge on a wide range of interests. Contingency plans had to be drawn up, identifying which organizations were responsible for deploying observers in the field and communicating their findings to a central agency. There had to be co-ordination in both a vertical and lateral sense, with a two-way flow of information both within and between organizations. NERC believed it was the obvious body to act as the central agency, relaying information to Government, co-ordinating the response to crises affecting the natural environment, and promoting longer-term surveillance and research programmes.[24]

The seabird wreck had also highlighted the value of birds as 'pretty critical indicators' of pollution. Citing the work at Monks Wood and elsewhere, Prestt (1971) described how the techniques developed for analysing organochlorine compounds in terrestrial and freshwater birds were equally applicable to seabirds. In its evidence to the Royal Commission on Environmental Pollution, NERC emphasized that a research council was ideally suited to carry out research to distinguish the effects of natural phenomena and pollution on birdlife, and the bearing of each on the population dynamics of as wide a range of species as possible.

NERC was under no illusions as to the difficulties that would arise in assuming the role of co-ordinator. The compilation of press statements and the eventual report on the seabird wreck had provided a foretaste of the problems that were bound to occur. Reflecting the complexity of the incident, and the number of far-flung and disparate bodies involved in its investigation, there had been many complaints of lack of consultation and other oversights. It was rare for the findings of one laboratory or agency to match exactly those of another. NERC was particularly conscious that much of the field data came in ways that could not 'be regarded as satisfactory in professional scientific circles'. As Mr Beverton put it, 'we are really having to accept the eye-witness accounts of a mere handful of observers whose strict impartiality is obviously not taken for granted since they have an axe to grind, praiseworthy axe though it may be'.[25]

Any attempt to resolve these wider issues had to take account of the aspirations and scope for conflict within the Council and its component bodies. Not surprisingly, the Toxic Chemicals and Wild Life Section had plenty of ideas for strengthening its role. As early as November 1969, there were proposals for setting up a special section at Monks Wood to monitor the status of seabirds more closely; it would include a scientific officer, two experimental officers, and a scientific assistant.[26] Dr Holdgate insisted that much more thought should be given to the proposals. He had, in fact, already received an intimation of the opposition they were likely to arouse within the Conservancy. An officer in Scotland had written, complaining that:

Monks Wood is in danger of becoming a juggernaut, engulfing all the rest of the Conservancy's initiative. Pollution affects all the environment, but Monks Wood is not the only centre of knowledge, and must not be allowed to extend its tentacles to all environments, and to inhibit growth elsewhere.

Monks Wood was not, he emphasized, the logical place to establish research on marine pollution. Scotland would be 'the best place for the Conservancy to organize research on any aspect of ornithology because it had the best group of ornithologists in Britain'. There were three major marine/oceanographic laboratories in Scotland, as well as several universities with strong marine interests.[27]

There was also a risk of any proposals emanating from Monks Wood prejudicing what were clearly going to be difficult discussions for the Conservancy on the overall contribution of NERC to pollution research. As early as December 1969, an *ad hoc* meeting was convened at NERC headquarters to discuss how and where NERC should support further pollution research. A NERC Pollution Unit was proposed, which would 'act as a co-ordinating agency to develop scientific guidelines for "outstations" collecting the data'. It would process and store this data, and provide an advisory and information service for the 'users'. The implications of such a proposal had to be carefully appraised by the Conservancy, in view of the influence a NERC Pollution Unit might exert over the direction and resources of the Toxic Chemicals and Wild Life Section, and the Conservancy in general. By 1970, such considerations had become part of the larger question of the Conservancy's relationship to NERC. Until this was resolved, the scope for major initiatives in the field of pollution was likely to be constrained.[28]

12

PESTICIDES AND THE ENVIRONMENTAL REVOLUTION

This book has recounted how the threat of pesticides to wildlife in Britain was first perceived, and how there was a wide range of responses to that perceived threat. It focuses on the organochlorine compounds which, of all the insecticides used, caused the most damage to wildlife. The account is based on published information, and on what can be found in the files of the Nature Conservancy, both at its headquarters and at the Monks Wood Experimental Station. Although most of the relevant files survive, they do not of course record everything, and there are many gaps in the narrative. Where possible, each episode had been corroborated by talking to those personally involved in the events described.

The book is by no means the first attempt to put the wildlife/pesticide issue into an historical perspective. Dunlap (1981) wrote an account of the DDT controversy and of its wider implications in the United States, In his words, the DDT episode:

can tell us something about the nature of American support for science and the ways in which the social, economic, and political context in which scientists worked affected both the research they did and the ways in which their knowledge was or was not used in making policy.

Although there were profound differences in the ways in which pesticides were used and regulated in Britain, an historical study might be equally illuminating, and it is hoped this book will help prepare the way for the time when the records of other bodies, both official and voluntary, become available for study and analysis.

The environmental revolution

There was a curious paradox about the early 1970s. European Conservation Year was great success. Whether measured in terms of the rising membership of the voluntary conservation bodies, the involvement of more and more people in conservation activities, or the exposure of wildlife and conservation issues on television and in the press, magazines, and books, everything seemed to auger well for the future. A new era had begun—and yet there was a sense of unease and frustration among many conservationists. All was not what it seemed!

In a speech to the Australian Conservation Foundation in April 1970, the Duke of Edinburgh described the sudden and explosive concern for the environment as the most remarkable shift in public outlook since the war. There was suddenly 'a massive and passionate concern for everything in and to do with nature and the pollution of the environment'. The change in attitude warranted the term 'environmental revolution' and, in his speech, Prince Philip developed the analogy with a revolution further. Having been dismissed as cranks initially, revolutionaries received overwhelming support as the significance of their warnings came to be heeded. There followed an inevitable backlash, and gradually a compromise evolved. The most menacing phase of a revolution was the return to some form of apathy. Conservation was like freedom: it could only be maintained by constant vigilance. As Prince Philip remarked, conservation was not something which could be pursued one day, and forgotten the next (Prince Philip 1978).

What some conservationists found so puzzling was the extent to which these different stages in the environmental revolution were telescoped together. It seemed that the revolution would be over before it had achieved very much in the way of real improvement. Nowhere was this more evident than in the field of pesticide control. The need to curb pollution seemed to be widely accepted, and such terms as 'environmental contamination' had slipped into common parlance. The 'anti-environment backlash' had begun,[1] with counter-attacks on the rhetoric of doom and despair. For the more discerning, there were even signs of the movement to regulate pesticides and pollution running out of steam. It was this quick succession, indeed merging, of the different stages of the revolution that some conservationists found so alarming.

There was no doubt that the revolution had passed through the first stages of its short life. In May 1963, the Council for Nature had organized a National Nature Week. Over 46 000 people had attended an exhibition sponsored by the *Observer*. The Week had been given considerable publicity on radio and television, and the Post Office had issued special stamps. To take advantage of unprecedented public interest, the Council for Nature and the Nature Conservancy hastily convened a conference, under the presidency of the Duke of Edinburgh. Its theme was 'The Countryside in 1970'. A wide variety of organizations and interests was brought together, in many cases for the first time, and the reports of the 12 study groups formed at the conference provided the substance of a Second Study Conference held in 1965 ('Countryside in 1970' 1964, 1966). The climax came at the third and final conference, in October 1970. Prince Philip spoke of how 'we may not have had much direct influence on the countryside itself, but I believe we have had a direct influence on men's minds. People have definitely become aware of conservation as a major issue' ('Countryside in 1970', 1970).

At the governmental level, a more cohesive approach to the perception and management of the environment seemed to be emerging. In late 1969, the Ministries of Housing and Local Government, and of Transport, were brought together under a new Secretary of State for Local Government and Regional Planning. A year later, and under a Conservative Government, the Department of the Environment (DOE) was formed. Things were happening on the pollution front. One of the priorities of the Secretary of State in late 1969 was 'to go urgently into the question of environmental pollution in all its forms'. Two months later, the Prime Minister announced the completion of the study, and the appointment of 'a permanent central unit, composed mainly of scientists', to assist the Secretary of State in his co-ordinating role on environmental pollution. This came to be called the Central Unit on Environmental Pollution. The appointment of a standing Royal Commission on Environmental Pollution was also announced in 1970.[2]

At Monks Wood, it was decided to extend the 'Open Weekend' in 1970 to an 'Open Week' in order to cope with the number of visitors. Over 8 000 came, including 65 school parties. In January, the Station was visited by the Prime Minister, the Secretary of State, and the Chief Scientific Adviser to the Government, Sir Solly Zuckerman. Arising out of discussions on the 'Countryside in 1970' committee, plans were well advanced for a visit by the Prince of Wales. Intended as a 'learning and working' visit, the informal visit took place eventually in November 1970.[3]

If the stamp of approval is a photograph of the Heir to the Throne, or the Prime Minister, listening to descriptions of work taking place in the laboratory, the Toxic Chemicals and Wild Life Section collected enough stamps to fill an album in 1970. And yet, somehow, these were not enough. Within three years, Monks Wood had been given a new role, and the Toxic Chemicals and Wild Life Section was no more.

Many believed that the environmental revolution was triggered by the 'environmental disasters' of the 1960s. According to a model set out by Ashby (1978), such hazards are usually identified in one of two ways. A 'catastrophe' may occur, such as the sudden and large-scale loss of bird-life in the spring of 1960/61. Alternatively, the scientist may alight on the presence of a hazard, often as an incidental part of his research. This happened when scientists discovered that PCBs were much more widely dispersed in the environment than previously supposed. Because an informed public is not necessarily a responsive one, it may be necessary to dramatize the perceived hazard in order to alert public opinion to its full significance. Without the high drama of *Silent spring*, Lord Ashby suggested, a further period might have elapsed before action was taken to curb the dangers arising from pesticide use.

Timing is of critical importance. It is doubtful whether *Silent spring* would have made as great an impact in Britain even two or three years earlier. By 1962/63, the warnings had become sufficiently credible for the public to want to probe more deeply. Over the previous decade, cir- cumstantial evidence had pointed to a casual link between pesticides and losses in birdlife, and the public was beginning to look, in some cases consciously, for a book which crystallized and gave expression to the doubts and misgivings which this evidence had implanted in the mind. In this context, it would be wrong to be too dismissive of the campaign to regulate the spraying of roadside verges, the investigations of the Zucker- man Working Parties, the (Sanders) Research Study Group, the Joint Committee of the BTO and RSPB, and the representations of the Nature Conservancy. Without them, *Silent spring* might well have found the mind of the British public uniformed and therefore unresponsive to advocacy.

The longer-term effects of *Silent spring* are also difficult to discern. It is one thing to arouse concern and to instil a desire to do something; it is another to ensure that these feelings are exploited to good effect. The conservationist has to strike a delicate balance. If he is too melodramatic, he will encourage a backlash to develop. If he sounds too complacent, he may give the impression that the cause has been won. In pressing for fur- ther extensions to 'the partial ban' on specific pesticides, for instance, it was easy to be manoeuvred into extreme positions.

In his book, *The Doomsday syndrome,* John Maddox (who was editor of *Nature* between 1966 and 1973) complained that one of the most dis- tressing features of the debate on the environment was the way in which it was assumed to be an argument 'between far-sighted people with the interests of humanity at heart, and others who care not tuppence for the future'. Those who were not ardently for the preservation of the environ- ment were presumed to be against it (Maddox 1972). Members of the Toxic Chemicals and Wild Life Section were frequently invited to write articles or give talks in the confident expectation that they would casti- gate all types of pesticide use. On more than one occasion, Dr Moore protested that 'it is far too readily assumed that pesticide users and con- servationists are inevitabily irreconcilable enemies' (Moore 1966*c* and 1970*c*). The Section's approach was summarized by Dr Moriarty. Because pesticides were certain, he said, to be used in large quantities for the foreseeable future, it was the responsibility of everyone to assess how far they were required to control individual pest species, bearing in mind the cost to the farmer and the losses that might be inflicted on other spe- cies. In this overall effort, it was the role of the Conservancy to identify the scale and significance of the damage inflicted on wildlife (Moriarty 1972*b*).

The Section's approach was, however, too subtle and sophisticated for those critics who accorded the Conservancy a different and largely stereotyped role. There were those in the pesticides and farming industries who sought to undermine the credibility of the Section's work by insi... that it was coloured by the purely emotive campaign being waged ... modern farming. There were others who claimed the Conservancy lacked the courage of its convictions, and had not gone far enough in denouncing pesticides.

Few 'technical' problems had ever aroused so much attention in the press and other media. In America, the tactics of confrontation went beyond public education, propaganda, and lobbying, and led to a direct challenge in the courts. Things never went so far in Britain, yet hardly a week passed in the 1960s without someone drawing attention to the dangers of pesticide use (Moore 1969d). *Silent spring* became a good conversation piece, irrespective of whether the book had been read or not. Pesticides and pollution were the target of serious comment, and of humour and satire. One of the most penetrating cartoons on the theme appeared in *Punch* in March 1963 (Fig. 12.1).

Far from being an asset, the increasing publicity given to ecology and conservation could be a liability. As Dr Mellanby commented in the Monks Wood report for 1969–71:

many so-called ecologists, with no real expertise or experience, are trying to jump on to the environmental bandwagon. They know that publicity will be given to those who make extreme statements, generally of impending disaster, and that those who prefer to rely on scientific observations will enjoy less popularity.

"*This is the dog that bit the cat that killed the rat that ate the malt that came from the grain that Jack sprayed.*"

Fig. 12.1 A cartoon from *Punch*, 6 March 1963.

During the Irish Sea seabird wreck, the Conservancy and NERC went to considerable lengths to distance themselves from this other kind of ecologist. Exceptional steps were taken to demonstrate the probity of the scientific evidence put forward.

Yet these steps were not enough. What should have been the greatest strength of the movement to regulate the use of pesticides, namely its increasingly close association with the wider desire to regulate pollution and the misuse of natural resources, turned out in fact to be its greatest liability. By the end of the 1960s, pollution, in all its manifold aspects, dominated the 'whole field of conservation and ecology' (Johnson 1973). Very quickly, the exalted tone of environmental rhetoric began to give way to stridency. It was assumed that the worst would always happen. No account was taken of the possibility that 'social institutions and human aspirations' might conspire to solve even the most daunting problems (Maddox 1972).

Grim and detailed scenarios were painted (Hare 1980). It was predicted that the whale would be extinct by 1973, and all important animal life in the sea would have disappeared by 1979. As deadlines for doom and destruction came and went, and nothing happened, the standing of the entire conservation movement suffered. Although the personal rapport and the links forged between the different interests at the conferences of the 'Countryside in 1970' and European Conservation Year survived, the position was never entirely retrieved. Critics of the conservation movement were in no mood to accept assurances that the reputation of most ecologists and conservationists was still intact. The self-acclaimed experts on the environment had been proved wrong.

It was against this confused background that observers tried to interpret trends in pesticide use. One of the most striking features of the agrochemical industry was the length of the lead-times required to synthesise a new pesticide and to achieve the economic break-even point. In its report of 1979, the Royal Commission on Environmental Pollution stated that it took 5 to 8 years to develop a pesticide and that between £10 million and £15 million might be spent at 1976 values before a product was placed on the market. The rate of introduction of new pesticides had slowed down markedly (Fig. 12.2). In 1956, the synthesis of new pesticides led, on average, to one commercial pesticide being launched for every 1 000 tested. The ratio had fallen to one in 5 000 in 1967, and to one in 10 000 by 1978 (Royal Commission 1979).

How far were the restrictions imposed on the industry responsible for these trends? It was established that pre-clearance safety and environmental studies made up 40 to 60 per cent of the costs of developing a pesticide. A survey in 1976 revealed, however, that manufacturers did not find the regulations at all onerous. They carried out more tests than

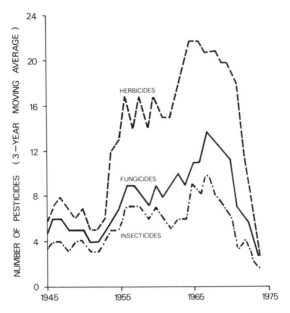

Fig. 12.2 Annual introduction of pesticides, 1945–75.

were required, largely because of the need to comply with the registration requirements of other countries (Tait 1976). Particularly at a time of economic recession, market forces may have been even more important in explaining the pattern discerned by the Royal Commission. For the first time in many years, pesticide sales actually fell in Britain during 1980.

Because of the previous success of the pesticides industry, Joyce Tait argued that it had reached a stage of 'technical maturity'. Manufacturers would have to look increasingly to the Third World for new and expanding markets (Tait 1981). The publication of the book, *A growing problem: pesticides and the Third World poor,* suggested that this was both happening and leading to serious problems of environmental contamination and human toxicity in those countries (Bull 1982). The British experience indicated perhaps that it was easy to exaggerate the effect of the professional conservationist, but hard to overestimate the value of his advice.

Turning to the way in which problems of pesticide control were perceived in agriculture, scientists in the Ministry of Agriculture found no evidence of the national survival of any species being threatened by pesticide use. When direct effects had been discerned, action had been taken to remedy the situation. The levels of surveillance over the direct effects of usage had generally been adequate, although Bunyan and Stanley

(1983) recommended further research on ways of defining and correcting the less direct effects on wildlife. It is doubtful, however, whether controls and monitoring programmes would have been introduced so quickly if it had not been for the advice and intervention of a third party in the self-regulatory system of voluntary regulation exercised by industry and agriculture. The nature conservation movement exerted a considerable, albeit indirect, influence over the timing and character of the British response to the chemical revolution on the land.

The success of the conservation movement was essentially negative in character. Species were preserved from destruction—a *status quo* was sought and, in most cases, attained. It was the same kind of negative success as the saving of a beautiful stretch of coastline from building development—its survival is soon taken for granted and the act of saving it is soon overlooked. It was not long before the pioneering role of the Nature Conservancy in the field of pesticide regulation was largely forgotten.

Nature's doctors

Whilst the shape, size, and endeavours of the Nature Conservancy do not constitute the story of the pesticide/wildlife debate, its existence counted for a very great deal. In her study of what happened in the state of Michigan, when fire retardent, containing the highly toxic PBB was mixed into animal feed, Joyce Egginton drew attention to the way in which the structure of government and its agencies could influence society's reponse to environmental problems. As one of the afflicted farmers rationalized, we have all the mistaken impression that government can take initiatives and swing into action when something new and unusual happens. Government cannot, however, run 'a private clinic for people's difficulties'. Only if you fall into an accepted group that is known to have problems, is there any chance of the whole system being mobilized on your behalf. That was the way bureaucracy worked (Egginton 1980). In the pesticides/wildlife debate in Britain, nature conservation had emerged from the 1940s as a 'recognized group', with a presence in government. Albeit extremely weak in relation to other groupings, there was at least somewhere to go in the corridors of government when seeking a redress for wildlife.

The Conservancy not only provided an entrée into Government, but it developed investigative powers of its own. Journalists dug deep into their bag of clichés to describe, and dramatize, the role of the scientist in the field of nature conservation. No one can recall who first used the label 'Natures doctors', but it was a label that gave rise to many discussions over coffee cups and elsewhere as to what were the limits of the doctors' practice, who should identify the patients, and where the balance should be struck between prevention and cure?

A scientist must make a powerful and eloquent case in promoting his work. If he does not, he will not be given the opportunity to carry out the research. In this sense, he has to be as political as any politician. In Lord Ashby's words:

the art of political leadership is to make proposals that are ahead, but only *just* ahead, of public opinion, and to time political action in such a way that it makes explicit what is already becoming implicit in the values held by the more articulate members of society. The password that the politician is waiting for is the word 'overdue'. When he hears that he knows it is becoming safe for him to act.

The scientist has to follow a similar course in securing the resources that enable him to perform his duties as a scientist.

Even if there were adequate resources for scientific endeavour, it is doubtful whether the scientist would be happy with the way in which these resources were deployed. In the real world, resources are never plentiful, and the scientist has to develop a faculty for influencing the way in which they are allocated. To do this, he has to persuade his paymaster that his work should be given priority over alternative channels of expenditure. In the case of the Nature Conservancy, this meant campaigning for a larger share of the science budget to be devoted to nature conservation. Officers in the Conservancy participated in committee work, and submitted evidence on the need for greater surveillance over pesticides, in the certain knowledge that if they succeeded in convincing others that pesticide research was a priority for investment, they would not only fulfil their role as official watchdogs over wildlife but their research budget would increase.

There was nothing inevitable about the Conservancy's ability to carry out research on toxic chemicals, let alone research that would win the respect of the scientific community. Progress in research owed much to expediency and the haphazard way in which the public and press, and more particularly the voluntary bodies and government agencies, came to perceive and respond to the alleged hazards. A great deal was left to the discretion of the individual scientist. Whilst there was a close rapport between members of the Toxic Chemicals and Wild Life Section, each saw his day-to-day research very much in personal terms. He decided his immediate goals and ways of achieving them. With hindsight, members of Monks Wood confess to being surprised at how well they read the situation, and how, by their individual efforts, they contributed to the overall message of the Conservancy at that time.

What made those 'informal chaotic personal interactions' so attuned to the times and so effective in their impact? In recalling the special excitement of working on pesticides at Monks Wood, Dr Ratcliffe paid tribute to the role of Norman Moore, without whose 'years of patient and wearing committee work within the bureaucracy, our boffin efforts would have counted for very little' (Ratcliffe 1980). Dr Moore was more than a buffer, protecting the creative scientist from the real world of public affairs. Having appointed his Section, he was its listening post and means of promoting its individual conclusions wherever they needed to be heard and understood. Although not a popularizer, he 'hammered the message home for ten years or more—where it really counts, among his colleagues and peers in science and agriculture'. Because his style and manner resembled that of the traditional university don, it came as a surprise to many to discover that he was 'discussing matters of urgent topical importance rather than a point of scholarship' (Chisholm 1972).

Despite the considerable interest aroused by the pesticide/wildlife issue, the resources devoted to it were always comparatively small. For several years, Robert Boote was not only the Conservancy's representative on the Advisory Committee, but he was the *only* officer on the administrative staff with any significant involvement in the question. His promotion of the conservation aspect, using every scrap of information that came his way, whether on committee or at a luncheon engagement, represented a remarkable *tour de force*. At Monks Wood, countless visitors came to see the campus devoted to pesticide research, and were astonished to find it all concentrated in a single corridor of a station with only a hundred staff. The number of senior scientific staff in the Toxic Chemicals and Wild Life Section never exceeded sixteen.

Whilst the comparative poverty of resources imposed severe constraints on what could be achieved, its significance should not be exaggerated during the critical decade of the 1960s when pesticide/wildlife research was still in a pioneer stage, and when it remained possible for the comparatively few to make a relatively large impact. That phase was, however, rapidly coming to an end in the early 1970s. Because of the great advances being made in this field or research in Britain and overseas, Dr Moore warned that the 'conservation usefulness' of the Section would have to be based in the future on much more 'intensive work on key problems'. This would require a commensurate increase in staff and resources. Unless this support was forthcoming, the Section would be faced with the choice of either spreading its resources too thinly or of giving up work in some fields in order to concentrate adequately on others.[4]

What the eventual outcome of this reorganization would have been, no one will know. Almost overnight, it became quite irrelevant. Developments both within and beyond the conservation movement led to the publication of a White Paper on the 'Framework of government research and development', which heralded a new era that was to have the most profound repercussions on pesticide/wildlife research and conservation generally.

As the Natural Environment Research Council (NERC) became fully established in the late 1960s, the Nature Conservancy came to feel the implications of being part of a larger research council. No longer did all the requests for guidance and comment on pesticides and other issues come directly from outside bodies to the Conservancy. The requests from Government had to be channelled through NERC. Instead of answering them directly, drafts were prepared for NERC, where they were scrutinized and perhaps modified by NERC personnel. This happened, for example, when the White Paper on pollution was being drafted in 1970 (see page 2). Not surprisingly, the NERC submission on the drafts was very different from that which the Conservancy would have made. Instead of focussing on the distinctive contribution of the Conservancy to pollution control, it took a corporate view of the different approaches and expertise contained in the Council's component bodies. In that sense, the Conservancy regarded the submission, and the White Paper itself, as a missed opportunity.[5]

Despite the marked advantages of being part of a larger Council, it became harder to promote the Conservancy as *the* official spokesman on nature conservation. The slowing down in the growth rate of the science budget provided further strain. Matters came to a head when structural changes were made in the way NERC was organized. The benefits of restoring the Conservancy's former independence were keenly debated and, in some quarters, advocated.

Meanwhile, in the spring of 1971, the Government had commissioned the Central Policy Review Staff (popularly known as the Think Tank) to review the whole range of government research and development, and particularly the way in which it was financed and organized. The Head of the Think Tank, Lord Rothschild, recommended that the funding of applied research and development should be placed on a customer-contractor basis. The appropriate government departments should themselves decide what type and level of research and development they required. In order to provide the departments with sufficient funds to act as customers in contracting the work, part of the research councils' budget would be transferred (Lord Privy Seal 1971).

The concept was endorsed by the Government in a White Paper of July 1972 (Lord Privy Seal 1972). This came as no surprise—the real bombshell for those in NERC was a further section of the White Paper, dealing *inter alia* with the Nature Conservancy. This described how the duality of functions in the Conservancy had led to 'stresses difficult to resolve within the present framework'. In the words of a government minister, the marriage of the conservancy girl into the research family, solemized by the Science and Technology Act of 1965, had not been a happy one.[6] If the original purposes of setting up the Conservancy in 1949, and NERC in 1965, were to be fulfilled, a fresh start had to be made. The Government had decided that the Conservancy's reserves, and the staff needed to run them, should become the responsibility of a new Nature Conservancy Council (NCC), which would be an independent statutory body, appointed by the Secretary of State for the Environment, and financed by a grant-in-aid. The funds for managing the reserves (about £1.1 million) and for commissioning applied research (about £0.3 million) would be transferred from NERC to the Department of the Environment (DOE). The balance of the Conservancy's budget (about £0.7 million), together with the stations and research staff, would remain in NERC.

In defending the decision to separate the conservation and research functions of the Conservancy, government ministers stressed that the Conservancy was only one of many agencies that had to be fitted into the infrastructure of government. As part of the much wider changes in the organization of government in 1970, the DOE had been created with the explicit purpose 'of placing in one Government Department the total approach to the environment'. It was only logical that the new department should include the official body responsible for nature conservation. By the same token, it was important to retain the research side of the Nature Conservancy in NERC, which had a complementary role to the DOE in the field of environmental research.[7]

A bill of divorce was needed. The 'split' and the abolition of the Nature Conservancy were enacted in a Nature Conservancy Council Bill, which was supported in principle by the Opposition and which received the Royal Assent in July 1973.[8] Vesting day for the Nature Conservancy Council was 1 November. Robert Boote became the new Director-General, with Ian Prestt (who had been on secondment to the Central Unit on Environmental Pollution) as his deputy. Derek Ratcliffe was appointed the Chief Scientist, and Norman Moore later became the Chief Advisory Officer. Three of Dr Ratcliffe's small team of scientists were recruited from the former Toxic Chemicals and Wild Life Section.

The future shape of the NERC component of the old Conservancy soon emerged. The Council of NERC decided to form 'a single organization closely knit in policy and adminstration', called the Institute of Terrestrial Ecology. With a staff of about 250, it would be one of the largest component bodies of NERC. Martin Holdgate became the new Director (he had left the Conservancy in 1970 to lead the Central Unit on Environmental Pollution). Jack Dempster became the head of a Division of Animal Ecology, and Officer-in-charge of Monks Wood, following the retirement of Professor Mellanby in August 1974. Under a completely new structure, the remaining personnel of the Toxic Chemicals and Wild Life Section were incorporated in the various sub-divisions. One of these was the Animal Function sub-division, made up of a small number of ecologists, physiologists, and a pharmacologist (Osborn 1980).

Having ennunciated and implemented the higher logic of government reorganization, it was left to others to discover and wrestle with the repucussions. The implications of the 'split' for research on pesticide use were highlighted when the Nature Conservancy Council (NCC) began to draft a submission to the Royal Commission on Environmental Pollution in 1977. At the time when the Royal Commission had been appointed in 1970, the Conservancy had included research staff working on toxic chemicals, and the Conservancy could speak with first-hand knowledge of the ecological and toxicological effects of organochlorine insecticides and PCBs. In 1977, and at a time when pollution problems continued to grow, there was no one engaged on pollution research in the NCC. In his evidence to the Royal Commission, the Chairman described how the Council was severely constrained in commissioning relevant research. Not only were there severe organizational difficulties in commissioning urgent and wide-ranging surveys and research, but there had been a general decline in the funds available. Whereas the former Conservancy had spent £400 000 per annum (at 1977 prices), the NCC had only £143 000 available to commission applied pollution research.

Canary in the coalmine

From the 1950s onwards, the perception of, and response to, the threat of pesticides to wildlife became an important theme in the conservation story. A vertical flow of information evolved, as reports of incidents, based on first-hand evidence, were relayed from the field to the headquarters of the voluntary bodies and government agencies. This information, together with the results of post-mortem or chemical analyses, was collated and assessed as it passed to the Advisory Committee, the minister's desk, or was taken up in parliament, the press, and elsewhere. A vertical structure, to some extent, developed in other countries, and on both a continental and international scale. In that sense, reports of the loss of breeding species or territory were not isolated incidents.

At no point in the vertical flow of intelligence were pesticides perceived in isolation. Much of the field evidence was acquired by those who worked on the land, or were well versed in natural history. The members of the Toxic Chemicals and Wild Life Section were appointed primarily for their wide ecological knowledge and only secondly for some relevant skill in pesticide research. For most people, their encounter with pesticides was only a part of a much wider concern for farming and wildlife. Pesticides were only one item on the agendas of meetings in the Conservancy and the voluntary bodies. Whilst the greater regulation of pesticides was desired, it was 'the whole system of modern agriculture and industry with its pressures on all types of land' which most endangered wildlife (Mellanby 1974). The files dealing with pesticides were only a few of the many that passed through the 'in trays' each day; they constitute only a small proportion of those which survive from the days of the Nature Conservancy.

Few problems were less amenable to generalization than the impact of each of the wide range of chemical pesticides on the different forms of wildlife. It was hardly surprising that there was so much room for widely divergent opinion, and yet, as Dr Moore commented, 'confessions of ignorance, whether made by government officials, industry or conservationists', were all too rare (Moore 1965e). It was in that context that so much importance was attached to the monthly meetings of the group of specialists, convened under the Pesticides Safety Precautions Scheme. Those who attended the meetings of the Inter-departmental Advisory Committee and the Scientific sub-committee were very much aware that they were representing specialized points of view, but that they were also expected to co-operate in reducing hazards to man and beast.

Likewise, a synoptic view was beginning to emerge as a result of the greater involvement of conservationists in the deliberations of such bodies as the British Crop Protection Council and ABMAC (later renamed the British Agrochemicals Association). Members of these various groupings became exposed to such concepts as integrated control and a concern for the environment, and the conservationist for his part learned much about the other man's preoccupations. As Dr Moore remarked, what started as a heated debate might evolve into effective action. This was no bad lesson to learn from the pesticides issue in European Conservation Year and afterwards.

For many, the pesticides issue provided them with their first involvement with research or conservation workers overseas, particularly after the NATO Advanced Study Institute at Monks Wood oin 1965. Dr Moore made 22 foreign trips between 1960 and 1974. He visited 11 different countries, and either met at Monks Wood or kept up a long correspondence with personnel in a further 13 countries. The benefits of the exchanges were mutual. Those at Monks Wood found the interest expressed in their work reassuring, particularly at the time of the *Further review* of organochlorines and in the period leading up to the 'split' of the Nature Conservancy. Sometimes, there was a tangible end-product—at the IUCN meeting in India in 1969, pressure exerted by Dr Moore and others helped to ensure that criticisms made of DDT continued to be based on scientific evidence rather than on emotions. At other times, there were only hints of the influence felt. A Russian correspondent asserted that the publications of the Toxic Chemicals and Wild Life Section had played a part in deterring that country from making large scale use of aldrin and dieldrin. There was no way of assessing whether this was true.

The rapport built up on the Advisory Committee and, more generally, under the aegis of the European Conservation Year, should not disguise the difficulties experienced by scientists, and those who sought to apply scientific knowledge. Disputes between scientists can be found in many fields. There is always likely to be controversy when the acceptance of one theory has the effect of displacing the theories held by others. Scientists are, after all, only human.

Ecologists were often surprised and disappointed by the lack of support given them by medical experts, when seeking extensions to the Pesticides Safety Precautions Scheme. In some instances, there may have been resentment at the way in which ecologists appeared to be trespassing on the medical field. In two papers, one entitled *Medicine and ecology*, Moore (1967*b*, 1969*c*) put forward another reason. He attributed the lack of sympathy to the fact that medicine was primarily concerned with the individual, whereas the ecologist's concern was generally for the population. Ecologists could, however, be equally compartmentalized in their approach. Some still held to the 'medieval' notion that man 'existed outside of nature', and that man-induced factors were irrelevant in understanding ecological processes. Such assumptions were tragic in view of the large part ecologists could play in controlling both pests *and* pesticides (Moore 1969*a*).

In a review written in 1967, Dr Moore described how the 'vigorous debate on the pros and cons of using certain pesticides has taught me something about the attitudes of people to Nature and to conservation, and has forced me to test the premises of arguments in favour of conservation' (Moore 1969*b*). The pesticides/wildlife debate was frequently described as a question of which was the more important—feeding the starving millions, or preserving the peregrine falcon? An editorial in *World Crops* spoke of how it was morally wrong 'to equate the possible destruction of a few predatory birds—embryonic life at that—with the lifetime of debilitating misery, once the destiny of millions of children born into malarial regions'. The introduction of DDT and other pesticides could be compared with the discovery of penicillin for the way in which it had improved the lot of Man (Anon. 1972*a*).

Moore (1964*c*) regarded this alleged choice between human misery and wildlife extinction as a sign of muddled thinking. The question of choice rarely arose in so stark a form—except in the minds of apologists for DDT. Where people were starving, wildlife conservation was completely secondary to food production, but even in these countries the role of wildlife as a food resource and as a focus for a dollar-earning tourist industry was coming to be recognized. Mellanby (1964) argued that, in Britain at least, 'we can afford not to squeeze the last penny out of our land if this is going to wipe out our wildlife'.

The pesticide issue highlighted the way in which the prevention or mitigation of man-made hazards was as much a political as a technical issue (Williams 1977). Britain had traditionally eschewed statutory controls wherever possible and had relied on a system of self-imposed regulation, the principal characteristics of which were flexibility, co-operation between all parties, and confidentiality (Gillespie 1979). It was an approach that reached its apotheosis in respect of air pollution (Ashby and Anderson 1981), but, as the experience of other countries indicated, it was not the only approach available. The response might no longer be adequate, either in meeting its declared goals or in the methods by which the objectives of pesticide control were attained.

The fact that a regulatory scheme might be voluntary did not imply that the public had an automatic right of access to the data being considered by the parties to that scheme, or to the proceedings that took place between those parties. In its report of 1979, the Royal Commission on Environmental Pollution conceded that some of the information submitted by pesticide manufacturers under the Pesticides Safety Precautions Scheme should remain confidential for commercial reasons, but it saw no technical reason why the Scheme should be generally so secretive. Wider publicity might enhance public confidence in the Scheme and the manner in which pesticides were used (Royal Commission 1979, and 1984). In its response to the Commission's report of 1979, the Government argued that 'a guarantee of confidentiality' was essential to the success of the control arrangements. It enabled the companies operating in a highly competitive industry to make available the full information required by the Advisory Committee on Pesticides for making 'proper safety assessments in the public interest' (Department of the Environment 1983).

In any assessment of how the controls over pesticides evolved, it is important to recognize how the doubts arising from the impact of pesticides on wildlife became both fused and confused with a wider and growing mistrust of science and technology. No matter how valuable they may be, all innovations are likely to have snags. For a time, pesticides became part of an almost obsessional reaction against all that was novel.

A leader in *Nature* described how DDT had become the focus of fashionable paranoia. It was an immense irony that a comparatively simple chemical, widely acclaimed as a great benefactor to mankind, should now become almost a symbol of all things fearsome (Anon. 1971*b*). The leader continued,

In some atavistic way, pillorying this chemical serves to satisfy a great many vague but common fears—that governments are incompetent in regulating the uses of innovation, that greedy manufacturers are urging farmers to use more of these materials than is strictly necssary, thus demonstrating yet another way in which the search for private profit will undermine society.

Perhaps the most strident attack on the motives of industry appeared in an American publication, *The pesticide conspiracy.* Drawing on his experience as an entomologist over 25 years, the author recounted how industry enticed farmers onto a 'pesticide treadmill', knowing that it offered no long term solutions to their pest problems. It was 'an ecological rip off' (Bosch 1980).

The growing disillusionment of the early 1960s stemmed not only from the polemical writings of *Silent spring,* but from the fact that instalments of *Silent spring* were published in the *New Yorker* at the same time as the nuclear test ban issue also reached a head. The thalidomide scandal was also about to break. As Dunlap (1981) commented, the image of contaminated rain, and of strontium 90 passing through grass into the milk of cows and nursing mothers, and into the next generation of children, did more than anything else 'to make Americans suspicious of the utopian dreams of technology'.

Many were the parallels drawn between the pesticides issue and the clamour that arose following the discovery of the injuries inflicted by the anti-depressant drug, thalidomide, on unborn babies (Insight Team 1979). Whilst it was misleading to draw any close comparisons, there were some similarities in the way the issues were perceived, enacted, and recorded. The thalidomide disaster, and the losses in wildlife arising from pesticide use, were no mere mischance. They resulted from the specific action of specific groups of people pursuing legitimate goals. There was nothing criminal in what they did—nothing for the police. With hindsight, some commentators have tried to make a virtue out of the incidents by suggesting that they alerted the world to unperceived dangers. This is misleading. The ability of drugs to cross the placental barrier and affect the foetus was already known and, likewise, the potential, if not the precise, dangers posed by pesticides to wildlife were widely understood. As early as 1945, the eminent insect physiologist, V.B. Wigglesworth, had warned readers of the *Atlantic Monthly* of the potential impact of DDT on the 'balance of nature'. In these senses, the disasters could have been avoided, or at least reduced in their impact, if different attitudes and procedures had prevailed.

In view of this wider concern for the human and natural environments, it would be misleading to suggest that the outcry against DDT and other pesticides was entirely in response to the harm inflicted on wildlife. The over-riding consideration was human welfare. In Britain and elsewhere, the fate of wildlife took on a new significance. Not for the first time, the Duke of Edinburgh drew an analogy with the canary in the coalmine. Whilst speaking at a Parliamentary Luncheon in Canberra in 1973, he recalled:

years ago, it was the custom of miners to take a caged canary down the mine. It was not to listen to its singing, but simply to warn them in case the air began to get foul. If the bird looked a bit pale and sick, it was time to watch out. If it fell off its perch, it was time to get out.

In the same way, the decline in the population of animal species, and their extinction in some cases, had first alerted the world to the changes being wrought by man. A concern for wildlife and wilderness areas was widening into an anxiety over the effects of these changes on the human environment itself (Prince Philip 1978).

Without such fears, it is extremely unlikely that the Government and parliament would have felt justified in taking steps to curb the use of pesticides simply to protect the peregrine falcon and other threatened species. It was, however, not only wildlife that died from having ingested lead arsenite sprayed on potato haulm in the late 1950; cattle and human beings had also perished. There was a real possibility of human beings consuming gamebirds, whose crops had been full of dressed corn. It was not only the golden eagle that ate the animals washed in the dieldrin-dips of the Highlands. It was the achievement of the nature conservation movement that this sense of unease over human welfare was used to promote the further study and protection of wildlife.

The word 'crisis' was increasingly used to describe the plight of wildlife and the environment generally. Terminology became confused. As Lord Ashby pointed out, a crisis is something that will pass, whereas the problems posed by environmental pollution, population growth, and the diminution of natural resources, were not of that type. They represented not so much a crisis as something which would remain with mankind for at least the foreseeable future (Ashby 1978). The regulation of pesticides was a significant pointer to what might happen more generally in the field of pollution. There was no chance of man being able to revert to a non-pesticide age; the task was instead to extract the advantages of this new aid to farming, without suffering from the drawbacks. As the full implications of persistence and the extent of contamination became clear, the challenge of mitigating the effects of pollutants came to be taken more seriously. An even greater premium had to be placed on forecasting the effects before they occurred, lest some might prove both harmful and irreversible.

In the journal of the Food and Agriculture Organization (FAO), Dr Moore wrote that we are living in a unique and critical period of time, when many of the problems are so new that history can provide no guidance in resolving them (Moore 1970b). Previously, most of the problems affecting the environment had arisen directly from some local event—perhaps the erection of a factory or drainage of a marsh. The response had been on a commensurate scale. The growth of the County Naturalists' Trusts had been a good example of this kind of response. As the pesticides issue demonstrated, an increasing number of problems were now finding expression considerable distances from where they were generated. Whether it was the application of a persistent pesticide or the wreck of an oil tanker, the repercussions might be felt on a regional, national, or even international scale (Prestt 1969a).

The last three decades have seen a significant change in the public conscience towards the natural environment. No longer is it enough to protect man from natural hazards; the environment must also be protected from the more hazardous activities of man. In an historical context, the shift in outlook is remarkable. The impact on wildlife, the scientific endeavour, and the pesticide story as a whole, were no isolated incident. Together, they provide an important case study in how the environmental revolution came about.

NOTES

Chapter 2

1 NCC, N32/8, Vol. 1
2 NCC, N32/8, Vol. 1
3 NCC, N32/8, Vol. 1
4 NCC, F/122; F/138; N32/8, Vol. 2
5 NCC, N/32/8, Vol. 2
6 Hansard, Commons, **530**, 8; NCC, N32/8, Vol. 4
7 NCC, N32/8, Vols. 4, 5
8 NCC, N32/8, Vol. 4
9 NCC, SP/M/55/40
10 NCC, N32/8, Vol. 5
11 NCC, N32/8, Vols. 5, 6.
12 NCC, N32/8, Vol. 6
13 NCC, N32/8, Vol. 6
14 NCC, N32/8A
15 NCC, N32/8, Vol. 6
16 NCC, N32/8, Vol. 7
17 NCC, N32/8, Vol. 8
18 NCC, N32/8B

Chapter 3

1 Public Record Office, MAF 132,1; Hansard, Lords, **176**, 403–25
2 NCC, N32, Vol. 1
3 NCC, N32, Vol. 1; Hansard, Commons **511**, 1448–9
4 NCC, N32, Vol. 1
5 NCC, N32/1
6 NCC, N32, Vol. 1; NC/Min/53/1
7 NCC, N32/2
8 NCC, N32/1–2, 4
9 NCC, N32, Vol. 1
10 NCC, N32/4
11 NCC, N32/3, Vol. 1
12 Hansard, Commons, **658**, *16–7*
13 NCC, N32/3

14 NCC, N32/3C
15 NCC, N32/9
16 NCC, NC/Min/58/4; N32, Vol. 2
17 NCC, N32/9
18 NCC, N32/8, Vol. 6
19 NCC, N32/9
20 Hansard, **590**, 1574–5
21 NCC, SP/M/58/58; N32/16, Vol. 1
22 NCC, SP/Min/59/1; SP/M/59/19
23 NCC, N32/3, Vol. 1
24 ITE, 52/3/H1

Chapter 4

1 *Farmers Weekly,* 17/11/58; NCC N32/15
2 NCC, F242/2, Vols. 1, 2
3 NCC, N7/27, Vol. 2; SP/M/59/7
4 NCC, SP/M/59/6
5 NCC, F242, Vol. 2; R16, Vol. 1
6 NCC, SP/M/59/13; N7/1A, Vol. 1; R16, Vol. 1
7 NCC, C120/2; NC/Min/59/2
8 NCC, N7/27, Vol. 2
9 NCC, SP/M/59/19
10 NCC, N32/10
11 NCC, SP/M/59/28
12 NCC, SP/Conf/59/5; F/Min/59/1; R16, Vol. 1
13 NCC, SP/M/59/40
14 NCC, N7/27, Vols, 2, 3
15 NCC, NC/Min/59/4, N7; R16, Vol. 1
16 NCC, R16, Vol. 1
17 NCC, R16, Vol. 1
18 NCC, SP/Min/60/1; SP/Conf/60/2
19 NCC, R16, Vol. 2; NC/Conf/60/1
20 NCC, SP/Min/60/2; NC/Min/60/2
21 NCC, R16, Vol. 2
22 NCC, R16A
23 NCC, P74/3–4; R16/3
24 NCC, R16/2/1, Vol. 1
25 NCC, NC/Min/60/4; NC/M/61/8
26 NCC, R16, Vol. 1
27 NCC, N32/10
28 NCC, N32/21; N32/16, Vol. 1
29 NCC, N32/18
30 NCC, N32/16, Vol. 1
31 Hansard, Lords, **219**, 749–52
32 Hansard, Commons, **613**, 1568–85
33 NCC, N32, Vol. 2
34 NCC, N32, Vol. 2
35 NCC, N32/10

36 ITE, 52/3/F4; NCC, N32/17
37 ITE, 52/3/F4
38 NCC, N32/3A, Vol. 2; ITE, 52/3/F2/1
39 ITE, 52/3/H1/1
40 ITE, 52/3/F2/1
41 NCC, N32/10; ITE, 52/3/D
42 ITE, 52/3/F4; 52/3/H11; NCC, N32/17
43 ITE, 52/3/H5
44 NCC, N32/10
45 NCC, N32/21

Chapter 5

1 NCC, N32/3, Vol. 1
2 NCC, N32/3, Vol. 1
3 NCC, N32/3A, Vol. 1
4 NCC, N32/16, Vol. 1; 1A17/02/06
5 NCC, N32/11; N32/16, Vol. 1
6 NCC, NC/Min/60/2; SP/M/60/29
7 NCC, S/50
8 Hansard, Commons, **617**, 648–50; *The Observer*, 27/3/60
9 *The Times*, 14/10/60
10 NCC, E/Min/60/4; ITE, 52/3/H1/1
11 NCC, NC/M/60/36; NC/Min/60/4; N32/16, Vol. 1
12 NCC, N32/16, Vol. 1
13 NCC, N32/3, Vol. 2
14 ITE, 52/3/H1/1
15 Hansard, Lords, **226**, 405–76; Lords, **229**, 509–16; Lords, **230**, 807–38;
16 NCC, N32/19
17 NCC, N32/19
18 ITE, 52/3/F2/1; 52/3/H1/1
19 NCC, N32/45
20 NCC, N32/16A; NC/M/61/48
21 NCC, N32/16A

Chapter 6

1 NCC, NC/32/16, Vol. 1
2 NCC, N32/23
3 ITE, 52/3/G5
4 NCC, N32/3/E (1A17/02/06)
5 NCC, N32/8, Vol. 6
6 NCC, N32/29, Vol. 1
7 NCC, N32/29, Vol. 1; N32/33
8 NCC, N32/29, Vol. 1
9 NCC, N32/29, Vol. 1
10 ITE, 52/3/F2/1
11 NCC, N32/29, Vol. 3

12 NCC, N32/29, Vol. 2
13 NCC, N32/29, Vol. 3; NC/M/63/5
14 ITE, 52/3/H7
15 Hansard, Lords, **247**, 1118–220
16 NCC, N32, Vol. 3
17 NCC, N32/38
18 ITE, 52/3/H7
19 NCC, SP/M/63/48, Annexes C & D
20 ITE, 52/3/F7/1
21 NCC, N32/17
22 NCC, N32/27
23 NCC, N32/26
24 NCC, N32/27
25 NCC, N32/36
26 ITE, 52/3/F2/1; 52/3/H1/1
27 NCC, N32/27
28 NCC, N32/26A; N32/33; ITE, 52/3/H5A
29 NCC, SP/M/62/36
30 NCC, N32/33; P74/6
31 NCC, R16/5
32 NCC, SP/Min/62/2; F212; R16/2/1, Vol. 1
33 NCC, R16/16

Chapter 7

 1 ITE, 52/3/F2/1
 2 NCC, N32/35
 3 ITE, 52/3/F2/1; NCC, N32/37
 4 NCC, N32/16A; N32/37–8
 5 Hansard, Lords, **250**, 958–61
 6 ITE, 52/3/G5
 7 Hansard, Lords, **250**, 1101; Hansard, Lords, **250**, 1277–82
 8 Hansard, Commons, **679**, 429–52
 9 ITE, 52/3/H7
10 NCC, SP/M/63/53, Appendix II
11 NCC, N32/40
12 NCC, N32/38; ITE, 52/3/F2/1
13 NCC, N32, Vol. 4
14 ITE, 52/8/1
15 NCC, F494/3; SP/M/60/13
16 NCC, SP/CONF/63/1
17 NCC, N32/16, Vol. 1
18 ITE, 52/3/G1
19 ITE, 52/3/H5
20 ITE, 52/3/A1
21 ITE, 52/8/2–3
22 ITE, 52/3/D
23 ITE, 52/3/G5

24 ITE, 52/3/F2/1
25 Hansard, Commons, **692**, 244–52; Lords, **256**, 1151–9
26 NCC, N32/40
27 Hansard, Commons, **692**, 756–70
28 NCC, N32/3A, Vol. 8; N32/40

Chapter 8

1 NCC, N32/4
2 NCC, N32/41
3 NCC, NC/M/70/117
4 NCC, N32/70 (1A17/06/02)
5 NCC, N32/4
6 NCC, N32/22 (1A17/07/03)
7 Kent Record Office, S/KR/AMp 2/12 and 15; Kent River Authority, Land Drainage Committee, minutes, 1966–8, Vol. 2
8 NCC, N32/51
9 NCC, N32/22 (1A17/07/03)
10 NCC, N32/51/1 (1A17/06/08)
11 NCC, N32/51
12 NCC, N32/22; N32/33/1
13 Hansard, Commons, **638**, 219; NCC, N32/20 (C10/05/09)
14 NCC, N32/20 (C10/05/09)
15 NCC, N7/36
16 NCC, C10/05/04
17 NCC, N32/20 (C10/05/09)

Chapter 9

1 NCC, N32/49
2 ITE, 52/3/K4/1
3 ITE, 52/3/A1
4 ITE, 52/3/G1; NCC, 09/3
5 NCC, N32/16, Vol. 2
6 ITE, 52/3/H6
7 ITE, 52/3/F8
8 Hansard, Lords, **304**, 124–5
9 NCC, N32/66; *The Observer*, 9/11/69
10 NCC, N7/1/N, Vol. 1–2
11 NCC, N7/1/N1
12 NCC, N7/1/N, Vol. 4
13 NCC, N7/1/N, Vol. 3
14 Hansard, Commons, **793**, 1351–5
15 NCC, N7/1/N, Vol. 5

Chapter 10

1 Hansard, Commons, **644**, 1695–724
2 NCC, NC/M/58/30, 32, 47
3 NCC, N32/21
4 NCC, N32/46
5 NCC, E/Min/68/1
6 NCC, SP/M/62/25
7 NCC, N32, Vol. 4; N32/26C; ITE, 52/3/F8/1–2
8 NCC, N32/26D
9 ITE, 52/3/K4/1
10 NCC, N7/1/E
11 NCC, N32, Vol. 3; N32/5
12 NCC, C10/10/07
13 NCC, N32/25; N32/65, Vol. 1
14 NCC, NC/Min/63/3
15 NCC, N32/40A
16 Hansard, Commons, **687**, *34–6*
17 Hansard, Commons, **689**, 333–46
18 Hansard, Commons, **688**, 820–6
19 NCC, N32/40A
20 NCC, N32/40A
21 Hansard, Lords, **255**, 1089–97
22 NCC, N32/35
23 NCC, N32/3/H1
24 NCC, N32/40
25 NCC, N32/3/H1
26 NCC, 1A17/01/02
27 Hansard, Lords, **263**, 397–90
28 NCC, N32/53B, Vol. 1
29 NCC, NC/M/68/194; N32/3/H1
30 *Nature, London,* **219**, 783
31 NCC, 1A17/05/03
32 NCC, N32/3/H1; ITE, 52/3/F2/3
33 NCC, N32/3/H1
34 ITE, 52/3/F2/4
35 Hansard, Commons, **834**, *172*
36 NCC, N32/72 (1A17/01/05)
37 NCC, N32/47, 61
38 NCC, N32/33/1
39 NCC, N32/33
40 NCC, N32/33/1
41 NCC, N32/73 (1A17/06/13); C10/10/10

Chapter 11

1 *The Field,* 17/12/64
2 NCC, N7/27A
3 NCC, F212

 4 NERC, D/1/7/27/2
 5 ITE, 52/3/B23
 6 NCC, N32/68 (1A17/06/01)
 7 NCC, N348/4, Vol. 1 (C1/04/05)
 8 NCC, N348/7 (C1/04/06)
 9 NCC, N348/6 (C1/04/07)
10 NCC, N348/1 (C1/04/05)
11 NCC, N348/5, Vol. 1 (C1/04/05)
12 NERC, CM/Min 69/7
13 NCC, N348/1 (C1/04/05)
14 NCC, N348/3 (C1/04/02)
15 NCC, N348/2 (C1/04/04)
16 NERC, D1/7/27/1, 4
17 NCC, N348/1 (C10/04/05)
18 NCC, N348/4, Vol. 1
19 NERC, D1/7/27/2
20 NCC, N348/3 (C1/04/04)
21 NCC, N348/2 (C1/04/04)
22 NERC, D1/7/27/3
23 NCC, N32/38; ITE, 52/3/H7
24 NCC, N348/1 (C1/04/05)
25 NCC, N348/4, Vol. 1
26 NCC, N348/2 (C1/04/04)
27 NCC. N348/4. Vol. 2
28 NCC, N348/1 (C1/04/05)

Chapter 12

 1 NCC, N32/33/1
 2 Hansard, Commons, **788**, 33–4; Commons, **793**, 638–45
 3 NCC, R16/28
 4 NCC, N32/33
 5 NCC, 232/10 (C5/16/04); N232/5 (C5/01/01)
 6 Hansard, Commons, **859**, 457
 7 Hansard, Commons **859**, 460–1
 8 Nature Conservancy Council Act, 1973, c.54

REFERENCES

ABMAC (1963). *Silent spring: a review and commentary*. Association of British Manufacturers of Agricultural Chemicals, London.

Advisory Committee on Pesticides and other Toxic Substances (1967). *Review of the present safety arrangements for the use of toxic chemicals in agriculture and food storage*. HMSO, London.

—— (1969). *Further review of certain persistent organochlorine pesticides used in Great Britain*. HMSO, London.

Advisory Committee on Poisonous Substances used in Agriculture and Food Storage (1964). *Review of the persistent organochlorine pesticides*. HMSO, London.

Agricultural Research Council (1964). *Report of the Research Committee on Toxic Chemicals*. Agricultural Research Council, London.

—— (1965). *Supplementary report of the Research Committee on Toxic Chemicals*. Agricultural Research Council, London.

Anderson, D. W., Hickey, J. J., Risebrough, R. W., Hughes, D. F. and Christensen, R. E. (1969). Significance of chlorinated hydrocarbon residues to breeding pelicans and cormorants. *Can. Fld Nat.* **83**, 91–112.

Anonymous (ed.) (1950) *Proceedings and papers of the International Technical Conference on the Protection of Nature, Lake Success, 1949*. International Union for the Protection of Nature, Brussels, 77–87.

—— (ed.) (1956). *Proceedings and papers of the 5th Technical Meeting*. International Union for the Protection of Nature, Brussels.

—— (1961). Notes and views. *Nature, Lond.* **190**, 488–9.

—— (1966). Report of a new chemical hazard. *New Scient.* **32**, 612.

—— (1969). DDT—uneasy compromise. *New Scient.* **44**, 626.

—— (1971*a*). Borlaug's warning. *Nature, Lond.* **233**, 444.

—— (1971*b*). DDT may be good for people. *Nature, Lond.* **233**, 437–8.

—— (1972). Editorial comment. *Wld Crops* **24**, 233.

—— (1976). *The history of pest control*. Fisons, Cambridge.

Ashby, E. (1978). *Reconciling man with the environment*. University Press, Oxford.

—— and Anderson, M. (1981). *The politics of clean air*. Clarendon Press, Oxford.

Bailey, S., Bunyan, P. J., Hamilton, G. A., Jennings, D. M., and Stanley, P. I. (1972). Accidental poisoning of wild geese in Perthsire, November 1971. *Wildfowl* **23**, 88–91.

Balme, O. E. (1954). Preliminary experiments on the effect of the selective weedkiller 2,4-D on the vegetation of roadside verges.. *Proc. Br. Weed Control Conf.* 1st, 219–28.

—— (1956). Conclusions of experiments on the effects of the selective weedkiller 2,4-D on the vegetation of roadside verges, *Proc. Br. Weed Control conf.* 2nd, 771–8.

Blackmore, D. K. (1963). The toxicity of some chlorinated hydrocarbon insecticides to British wild foxes. *J. comp. Path. Ther.* **73**, 391–409.

Borlaug, N. E. (1972). Mankind and civilization at another crossroad. In balance with nature—a biological myth. *Bio Sci.* **22**, 41–4.

253

Bosch, R. van den (1980). *The pesticide conspriracy.* Prism Press, Dorchester.

Bull, D. (1982). *A growing problem: pesticides and the Third World poor.* Oxfam, Oxford.

Bunyan, P. J. and Stanley, P. I. (1983). The environmental cost of pesticide usage in the United Kingdom. *Agric. Ecosyst. Environ.* **9**, 187–209.

Cabinet Office (1967). *The Torrey Canyon.* HMSO, London.

Cairns, T. and Siegmund, E. G. (1981). PCBs. Regulatory history and analytical problems. *Analyt. Chem.* **53**, 1183–93.

Campbell, B. (1961). Review. *Br Birds.* **54**, 164–6.

Carnaghan, R. B. A., and Blaxland, J. D. (1957). The toxic effects of certain seed-dressings on wild and game birds. *Vet. Rec.* **69**, 324–5.

Carson, R. (1962). Silent Spring. *New Yorker* (no volume number) 16 June, 35–99, 23 June, 31–89, and 30 June, 35–67.

—— (1963). *Silent spring.* Hamish Hamilton, London.

Chadwick, G. G. and Brocksen, R. W. (1969). Accumulation of dieldrin by fish and selected fish-food organisms. *J. Wildl. Mgmt.* **33**, 693–700.

Chisholm, A. (1972). *Philosophers of the earth.* Sidgwick & Jackson, London, 149–57.

Cooke, A. S. (1970). The effect of pp'-DDT on tadpoles of the common frog. *Environ. Pollut.* **1**, 57–71.

—— 1 (1971). Selective predation by newts on frog tadpoles treated with DDT. *Nature, Lond.* **229**, 275–6.

—— (1973). Shell thinning in avian eggs by environmental pollutants. *Environ. Pollut.* **4**, 85–152.

—— (1975). Pesticides and eggshell formation. *Symp. zool. Soc. Lond.* **35**, 339–61.

—— (1979). Changes in egg shell characteristics of the sparrowhawk and peregrine associated with exposure to environmental pollutants during recent decades. *J. Zool., Lond.* **187**, 245–63.

Cornwallis, R. K. (1963). Birds of prey conference. *Bird Study* **10**, 44.

Countryside in 1970 (1964). *Proceedings of the Study Conference.* HMSO, London.

—— (1966). *Second conference.* Royal Society of Arts, London.

—— (1970). *Third conference.* Royal Society of Arts, London.

Cramp, S. (1963). Toxic chemicals and birds of prey. *Br Birds* **56**, 124–39.

—— (1969). Control of DDT urgently needed. *Birds* **2**, 264.

Cramp, S. and Conder, P. J. (1961). *The deaths of birds and mammals.* Royal Society for the Protection of Birds, London.

—— and Ash, J. S. (1962) *Deaths of birds and mammals from toxic chemicals.* Royal Society for the Protection of Birds, London.

—— (1963). *deaths of birds and mammals from toxic chemicals.* Royal Society for the Protection of Birds, Sandy.

Davis, B. N. K. (1964*a*). Toxic chemicals and soil animals. In *Food supply and nature conservation* (ed. Anonymous), pp. 15–24. Cambridgeshire College of Arts and Technology, Cambridge.

—— (1964*b*). Review, *J. Ecol.* **52**, 447–8.

—— (1965). The immediate and long-term effects of the herbicide MCPA on soil arthropods. *Bull. ent. Res.* **56**, 357–66.

—— (1966). Soil animals as vectors of organochlorine insecticides for ground feeding birds. *J. appl. Ecol.* 3 Suppl., 133–9.

—— (1967). Bird feeding preferences among different crops in an area near Huntingdon. *Bird Study* **14**, 227–37.

—— (1968). The soil macrofauna and organochlorine insecticide residues at twelve agricultural sites near Huntingdon. *Ann. appl. Biol.* **61**, 29–45.

—— (1974). Levels of dieldrin in dressed wheat seed after drilling and exposure on the soil surface. *Environ. Pollut.* **7**, 309–17.

—— and French, M. C. (1969). The accumulation and loss of organochlorine insecticide residues by beetles, worms and slugs in sprayed fields. *Soil Biol. Biochem.* **1**, 45–55.

—— and Harrison, R. B. (1966). Organochlorine insecticide residues in soil invertebrates. *Nature, Lond.* **211**, 1424–5.

Dempster, J. P. (1967). The control of *Pieris rapae* with DDT. I. The natural mortality of the young stages of *Pieris., J. appl. Ecol.* **4**, 485–500.

—— (1968*a*). The control of *Pieris rapae* with DDT. II. Survival of the young stages of *Pieris* after spraying. *J. appl. Ecol.* **5**, 451–62.

—— (1968*b*). The control of *Pieris rapae* with DDT. III. Some changes in the crop fauna. *J. appl. Ecol.* **5**, 463–75.

—— (1968*c*). The sublethal effect of DDT on the rate of feeding by the ground-beetle *Harpalus rufipes*. *Entomologia exp. appl.* **11**, 51–4.

—— (1975). *Animal population ecology*. Academic Press, London.

Department of the Environment (1974*a*). The monitoring of the environment in the United Kingdom. *Pollut. Paper* 1. HMSO. London.

—— (1974*b*). The non-agricultural uses of pesticides in Great Britain. *Pollut Paper* 3. HMSO, London.

—— (1983). Agriculture and pollution. *Pollut. Paper* 21, HMSO, London.

Dunlap, T. R. (1981). *DDT: scientists, citizens and public policy*. University Press, Princeton.

Edson, E. F. (1964). A review of chemical pest control methods. In *Food supply and nature conservation* (ed. Anonymous), pp. 35–8. Cambridgeshire College of Arts and Technology, Cambridge.

Egan, H. (1967). Pesticide quest: residue surveys and tolerances. *Chemy Ind.* 1721–30

Egginton, J. (1980). *Bitter harvest*. Secker and Warburg, London.

Elton, C. E. (1966). *The pattern of animal communities*. Methuen, London.

—— and Miller, R. S. (1954). The ecological survey of animal communities. *J. Ecol.* **42**, 460–96.

Forestry Commission (1970). *Forty ninth annual report*. HMSO, London.

Fryer, J. D. (1964). The herbicide revolution. *New Scient.* **24**, 522–4.

George, J. L. and Frear, D. E. H. (1965). Pesticides in the Antarctic. *J. appl. Ecol.* 3 Suppl., 155–67.

Gillespie, B. (1979). British 'safety policy' and pesticides. In *Directing technology: policies for promotion and control* (ed. R. Johnston and P. Gummett), pp. 202–24. Croom Helm, London.

Gough, H. C. (1957). Studies on wheat bulb fly. IV The distribution of damage in England and Wales in 1953. *Bull. ent. Res.* **48**, 447–57.

Greenberg, D. S. (1963). Pesticides. *Science, N. Y.* **140**, 878–9.

Gunn, D. L. (1972). Dilemmas in conservation for applied biologists. *Ann. appl. Biol.* **72**, 105–27.

Hailsham, Lord (1963). Effect of toxic chemicals on wild life. *Nature Lond.* **200**, 826–7.

—— (1975). *The door wherein I went*. Collins, London, 180–6.

Hamilton, G. A. and Stanley, P. I. (1975). Further cases of poisoning of wild geese by an organochlorine winter wheat seed treatment. *Wildfowl* **26**, 49–54.

Hare, F. K. (1980). The planetary environment: fragile or sturdy. *Geogrl J.* **146**, 379–80.

Heuval, M. J. van den (1975). The United Kingdom approach to the problems of assessing potential hazards to wildlife from the use of agricultural chemicals. In *European colloquium: problems raised by the contamination of man and his environment by persistent pesticides and organohalogenated compounds* (ed. Anonymous) pp. 293–300. Commission of the European Community, Luxembourg.

Hickey, J. J. (ed.) (1969). *Peregrine falcon populations: their biology and decline.* University Press, Wisconsin.

—— and Anderson, D. W. (1968). Chlorinated hydrocarbons and eggshell changes in raptorial and fish-eating birds. *Science, N.Y.* **162**, 271–3.

Holden, A. V. (1970). Source of polychlorinated biphenyl contamination in the marine environment. *Nature, Lond.* **228**, 1220–1.

Holdgate, M. W. (ed.) (1971). *The seabird wreck in the Irish Sea. Autumn 1969.* Natural Environment Research Council, London.

Holdich, B. (1825). *An essay on the weeds of agriculture.* Ridgway, London. 69–70.

Holmes, D. C., Simmons, J. H. and Tatton, J. O' G. (1967). Chlorinated hydrocarbons in British wildlife. *Nature, Lond.* **216**, 227–9.

Hunt, E. G. and Bischoff, A. I. (1960). Inimical effects on wildlife of periodic DDD applications to Clear Lake *California Fish & Game,* **46**, 91–106.

Insight Team of the Sunday Times (1979). *Suffer the children: the story of thalidomide.* Deutsch, London.

Jefferies, D. J. (1967). The delay in ovulation produced by pp'-DDT and its possible significance in the field. *Ibis* **109**, 266–72.

—— (1969). Causes of badger mortality in eastern counties in England. *J. Zool. Lond.* **157**, 429–36.

—— (1971). Some sublethal effects of pp'-DDT and its metabolite pp'-DDE on breeding passerine birds. *Meded. Fakult. Landbouw-wetenschappen Gent* **36**, 34–42.

—— (1972). Organochlorine insecticide residues in British bats and their significance. *J. Zool. Lond.* **166**, 245–263.

—— (1973). The effects of organochlorine insecticides and their metabolites on breeding birds. *J. Reprod. Fert.* **19**, Suppl., 337–52.

—— (1975). The role of the thyroid in the production of sublethal effects by organochlorine insecticides and polychlorinated biphenyls. In *Organochlorine insecticides: persistent organic pollutants* (ed. F. Moriarty), pp. 132–230. Academic Press, London.

—— and Davis, B. N. K. (1968). Dynamics of dieldrin in soil, earthworms, and song thrushes. *J. Wildl. Mgmt.* **32**, 441–56.

—— and French, M. C. (1969). Avian thyroid: effect of pp'-DDT on size and activity. *Science, N. Y.* **166**, 1278–80.

—— (1971). Hyper- and hypothyroidism in pigeons fed on DDT: an explanation for the 'thin eggshell phenomenon'. *Environ. Pollut.* **1**, 235–42.

—— (1972a). Changes induced in the pigeon thyroid by pp'-DDE and dieldrin. *J. Wildl. Mgmt.* **36**, 24–30.

—— (1972b). Lead concentrations in small mammals trapped on roadside verges and field sites. *Environ. Pollut.* **3**, 147–56.

—— (1976). Mercury, cadmium, zinc, copper and organochlorine insecticide levels in small mammals trapped in a wheat field. *Environ. Pollut.* **10**, 175–82.

—— and Prestt, I. (1966). Post-mortems of peregrines and lanners with particular reference to organochlorine residues. *Br. Birds* **59**, 49–64.

—— Stainsby, B. and French, M. C. (1973). The ecology of small mammals in arable fields drilled with winter wheat and the increase in their dieldrin and mercury residues. *J. Zool. Lond.* **171**, 513–39.

—— and Walker, C. H. (1966). Uptake of pp'-DDT and its post-mortem break-down in the avian liver. *Nature, Lond.* **212**, 533–4.

Jensen, S. (1972). The PCB story. *Ambio* **1**, 123–32.

—— Johnels, A. G., Olsson, M. and Otterlind, G. (1969). DDT and PCB in marine animals from Swedish waters. *Nature, Lond.* **224**, 247–50.

Johnson, S. P. (1973). *The politics of environment.* Tom Stacey, London.

Kinnier-Wilson, J. V. (1979). *The rebel lands.* University Press, Cambridge.

Kuenen, D. J. (ed.) (1961). *The ecological effects of biological and chemical control of undesirable plants and animals. Eighth Technical Meeting, 1960.* E. J. Brill, Leiden.

Laboratory of the Government Chemist (1962). *Report of the Government Chemist, 1961.* HMSO, London, 51–61.

—— (1964). *Report of the Government Chemist, 1963.* HMSO, London, 67–79.

—— (1974). *Report of the Government Chemist, 1973.* HMSO, London, 9–20.

Laverton, S. (1962). *The profitable use of farm chemicals.* Oxford University Press, London.

Lockie, J. D. and Ratcliffe, D. A. (1964). Insecticides and Scottish golden eagles. *Br. Birds* **57**, 89–101.

—— Ratcliffe, D. A., and Balharry, R. (1969). Breeding success and organochlorine residues in golden eagles in west Scotland. *J. Appl. Ecol.* **6**, 381–9.

—— Stephen, D. (1959). Eagles, lambs and land management on Lewis. *J. anim. Ecol.* **28**, 43–50.

Lord President of the Council and Minister for Science (1960). *Annual report of the Advisory Council on Scientific Policy, 1959–60,* Cmnd 1167.

—— (1963). *Annual report of the Advisory Council on Scientific Policy, 1962–1963,* Cmnd 2163.

Lord Privy Seal (1971). *A framework for government research and development.* Cmnd 4814.

—— Framework for government research and development. Cmnd 5046.

Lowery, G. H. (1969). An examination of a worldwide disaster. *Science, N.Y.* **166**, 591.

Macgregor, A. G. (ed.) (1964). Toxic effects of pest-control agents. *Adv. Sci.* **21**, 5–10.

Maddox, J. (1972). *The doomsday syndrome.* Macmillan, London.

Martin, H. (1963). Present safeguards in Great Britain against pesticide residues and hazards. *Residue Rev.,* **4**, 17–32.

Mellanby, K. (1963). Monks Wood Experimental Station. *Nature, Lond.* **200**, 825–6.

—— (1964). Pesticides and wildlife. *Proc. R. Instn Gt. Br.* **40**, 119–28.

—— (1967). *Pesticides and pollution.* Collins, London.

—— (1974). The future of biological control in Britain; a conservationist's view. In *Biology in pest and disease control* (ed. D. Price Jones and M. E. Solomon, pp. 349–53. Blackwell, British Ecological Society Symposium, 13, Oxford.

Metcalf, R. L. (1980). Changing role of insecticides in crop protection. *Ann. Rev. Entomology* **25**, 219–56.

Middleton, A. D. (1956). The effects of certain insecticides and herbicides on game-birds and other wildlife in farm crops. In *Proceedings and papers of the 5th Technical Meeting* (ed. Anonymous). pp. 161–3. International Union for the Protection of Nature, Brussels.

Miller, E. J. (1965). The Pesticides Safety Precautions Scheme. *Residue Rev.* **11**, 100–18.

Milstein, P. le S., Prestt, I. and Bell, A. A. (1970). The breeding cycle of the grey heron. *Ardea* **58**, 171–257.

Ministry of Agriculture, Fisheries and Food. (1963). *Chemicals for the gardener.* HMSO, London.

Moore, N. W. (1957). The past and present status of the buzzard in the British Isles. *Br. Birds* **50**, 173–97.

—— (1962*a*) Pesticides and wildlife, the conservationst's point of view. *Chemy. Ind.* 2130–1.

—— (1962*b*). Toxic chemicals and birds: the ecological background to conservation problems. *Br. Birds* **55**, 428–35.

—— (1964*a*). Man, pesticides and the conservation of wildlife. *Biology hum. Affairs* **29**, 19–25.

—— (1964*b*). Pesticides and wildlife. *Adv. Sci.* **21**, 5–10.

—— (1964*c*). Toxic chemicals and conservation. In *Food supply and nature conservation* (ed. Anonymous), pp. 13–5. Cambridgeshire College of Arts and Technology, Cambridge.

—— (1965*a*). Environmental contamination by pesticides. In *Ecology and the industrial society* (ed. G. T. Goodman), pp. 219–37. Blackwell, British Ecological Society Symposium, 5, Oxford.

—— (1965*b*). Food-production needs and environmental protection. *Munic. Engng* **3962**, 2583–4.

—— (1965*c*). Foreword. In *Pesticides and the living landscape* (by R. L. Rudd). Faber & Faber, London.

—— (1965*d*). Pesticides and birds—a review of the situation in Great Britain in 1965 *Bird Study* **12**, 222–52.

—— (1965*e*). Pesticides and the conservation of wildlife on farms and in gardens. *Jl R. agric. Soc.* **125**, 73–7.

—— (1966*a*). An assessment of the discussions of the Symposium on Pesticides in the Environment and their Effects on Wildlife. *J. appl. Ecol.* 3 Suppl., 291–5.

—— (1966*b*). A pesticide monitoring system with special reference to the selection of indicator species. *J. appl. Ecol.* 3 Suppl., 261–9.

—— (1966*c*). Insecticides and wildlife. *Proc. R. ent. Soc. Lond. (C)*, **31**, 21–2.

—— (1967*a*). Effects of pesticides on wildlife. *Proc. R. Soc, Lond. (B)* **167**, 128–33.

—— (1967*b*). Organochlorine insecticides and wildlife populations. *Proc, R. Soc. Med.* **60**, 23–4.

—— (1967*c*). A synopsis of the pesticide problem. *Adv. ecol. Res.* **4**, 75–129.

—— (1969*a*). Can hazards to wildlife be assessed by field trials before registration? *Proc. Br. Insectic. Fungic. Conf.*, 5th 699–703.

—— (1969b). Experience with pesticides and the theory of conservation. *Biol. Conserv.* **1**, 201–7.

—— (1969c). Medicine and ecology. *Rur. Med.* **1**, 27–8.

—— (1969d). The significance of the persistent organochlorine insecticides and the polychlorinated biphenyls. *Biologist* **16**, 157–62.

—— (1970a). The ecological impact of pollution. In *Papers and Proceedings of the 11th Technical Meeting* (ed. Anonymous). pp. 76–81. International Union for the Conservation of Nature, Morges.

—— (1970b). Implications of the pesticide age. *Ceres, FAO* **3**, 27–9.

—— (1970c). Pesticides and conservation. *Proc. Br. Weed Control Conf.* 10th, 1032–5.

—— (1970d). Pesticides know no frontiers. *New Scient.* **46**, 114–5.

—— (1971). Implications of success and failure in the control of pesticides. In *Environment, man, survival* (ed. L. H. Wullstein, I. B. McNulty and L. Klikoff, pp. 75–86. University Press, Grand Canyon Symposium, Utah.

—— (1973). Indicator species. *Nature in focus* **14**, 3–6.

—— (1975). Pesticide monitoring from the national and international points of view. *Agric. & Environ.* **2**, 75–83.

—— (1981) Concerted action: pesticides. In *The principles of pollution control: the British experience* (ed. The Committee for the Envrionment), pp. 31–7. Royal Society of Arts, London.

—— and Ratcliffe, D. A. (1962). Chlorinated hydrocarbon residues in the egg of a peregrine falcon from Perthshire. *Bird Study* **9**, 242–4.

—— and Tatton, J. O'G. (1965). Organochlorine insecticide residues in the eggs of sea birds. *Nature, Lond.* **207**, 42–3.

—— and Walker, C. H. (1964). Organic chlorine insecticide residues in wild birds. *nature. Lond.* **201**, 1072–3.

Moore, W. C. (1964). The Notification Scheme. In *Food supply and nature conservation* (ed. Anonymous), pp. 39–44. Cambridgeshire College of Arts and Technology, Cambridge.

Moriarty, F. (1968). The toxicity and sublethal effects of pp'-DDT and dieldrin to *Aglais urticae* and *Chorthippus brunneus*. *Ann. appl. Biol.* **62**, 371–93.

—— (1972a). The effects of pesticides on wildlife: exposure and residues. *Sci. total Environ.* **1**., 267–88.

—— (1972b). Pesticides and wildlife: retrospect and prospect. *OEPP/EPPO Bulletin* **4**, 51–63.

—— (1972c). Pollutants and food-chains. *New scient.* **53**, 594–6.

—— (1975a). The dispersal and persistence of pp'-DDT. In *The ecology of resource degradation and renewal* (ed. M. J. Chadwick and G. T. Goodman), pp. 31–47. Blackwell, British Ecological Society Symposium 15, Oxford.

—— (1975b). *Pollutants and animals: a factual perspective.* Allen & Unwin, London.

—— (1975c). Exposure and residues. In *Organochlorine insecticides: persistent organic pollutants* (ed. F. Moriarty), pp. 29–72. Academic Press, London.

—— and French, M. C. (1977). Mercury in waterways that drain into the Wash, in eastern England. *Water Res.* **11**, 367–72.

Murton, R. K. (1977). Pesticides and wildlife: current ITE research. In *Some aspects of research on pesticides* (ed. Inter-Research Council Committee on Pollution Research), pp. 11–8. Natural Environment Research Council, IRCCOPR Seminar Report 3, London.

—— and Vizoso, M. (1963). Dressed cereal seed as a hazard to wood-pigeons. *Ann. appl. biol.* **52**, 503–17.

Nature Conservancy (1969). *Progress, 1964–1968.* Nature Conservancy, London.

Nelson, B. (1969). Herbicides: order on 2,4,5-T issued at unusually high level. *Science, N.Y.* **166**, 977–9.

Newton, I. (1974). Changes attributed to pesticides in the nesting success of the sparrowhawk in Britain. *J. appl. Ecol.* **11**, 95–102.

—— and Bogan, J. (1974). Organochlorine residues, eggshell thinning and hatching success in British sparrowhawks. *Nature, Lond.* **249**, 582–3.

Nicholson, E. M. (1965). Advances in British nature conservation. *Handbk SPNR*, 1–16.

—— (1970), *The environmental revolution.* Hodder & Stoughton, London.

Osborne, D. (1980). Pesticides and British wildlife: brief review of some recent work at Monks Wood. *J. R. Soc. Med.* **73**, 127–30.

Parslow, J. L. F. (1970). Oil pollution and seabirds. NATO Committee on the challenges of modern society **1**, 1–12.

—— (1973). Mercury in waders from the Wash. *Environ Pollut.* **5**, 295–304.

—— and Jefferies, D. J. (1973). Relationship between organochlorine residues in livers and whole bodies of guillemots. *Environ Pollut.* **5**, 87–101.

——and French, M. C. (1972). Ingested pollutants in puffins and their eggs. *Bird Study* **19**, 18–33.

Peakall, D. B., Reynolds, L. M., and French, M. C. (1976). DDE in eggs of the peregrine falcon. *Bird Study* **23**, 183–6.

Pile, W. (1979). *The Department of Education and Science.* Allen & Unwin, London, 202–16.

Pollard, E. (1971). Hedges, VI. Habitat diversity and crop pests: a study of *Brevicoryne brassicae* and its syrphid predators. *J. appl. Ecol.* **8**, 751–80.

——and Relton, J. (1970). Hedges, V. A study of small mammals in hedges and cultivated fields. *J. appl. Ecol.* **7**, 549–57.

—— Hooper, M. D., and Moore, N. W. (1974). *Hedges.* Collins, London.

Potter, C. (1956). Work in England on effect of insecticides and other chemicals used in plant protection on beneficial insect and insect populations. In *Proceedings and papers of the 5th Technical Meeting (ed. Anonymous),* pp. 114–21. International Union for the Protection of nature, Brussels.

President's Science Advisory Committee (1963). *Use of pesticides.* White House, Washington.

Prestt, I. (1965). An enquiry into the recent breeding status of some of the smaller birds of prey and crows in Britain. *Bird Study* **12**, 196–221.

—— (1966). Studies of recent changes in the status of some birds of prey and fish-feeding birds in Britain. *J. appl. Ecol.* 3 Suppl., 107–12.

—— (1967). Investigations into possible effects of organochlorine insecticides on wild predatory birds. *Br. Insectic. Fungic. Conf., 4th* **1**, 26–35.

—— (1969*a*). A conservationist's approach to wildlife problems. *Naturalist, Hull* **910**, 73–6.

—— (1969*b*). Recent population trends among British raptors. In *Peregrine falcon populations: their biology and decline* (ed. J. J. Hickey), pp. 283–7. University Press, Wisconsin.

—— (1970). Organochlorine pollution of rivers and the heron. In *Papers and Proceedings of the 11th Technical Meeting* (ed. Anonymous), pp. 95–102. International Union for the Conservation of Nature, Morges.

—— (1971). Techniques for assessment of pollution effects on seabirds. *Proc. R. Soc. Lond.* (B). **177**, 287–94.

—— and Jefferies, D. J. (1969). Winter numbers, breeding success, and organochlorine residues in the great crested grebe in Britain. *Bird Study* **16**, 168–85.

—— and Jefferies, D. J. and Macdonald, J. W. (1968). Post-mortem examinations of four rough-legged buzzards. *Br Birds*. **61**, 457–65.

—— and Mills, D. H. (1966). A census of the great crested grebe in Britain, 1965. *Bird Study* **13**, 163–203.

—— and Ratcliffe, D. A. (1972). Effects of organochlorine insecticides on European birdlife. *Proc. Int. Orn. Congr.* **15**, 486–513.

—— and Moore, N. W. (1970). Polychlorinated biphenyls in wild birds in Britain and their avian toxicity. *Environ. Pollut.* **1**, 3–26.

Prime Minister (1963). *Committee of Enquiry into the Organization of the Civil Service,* Cmnd 2171.

Prince Philip (1978). *The environmental revolution: speeches on conservation, 1962–77.* Deutsch, London.

Ratcliffe, D. A. (1958). Broken eggs in peregrine eyries. *Br. Birds* **51**, 23–6.

—— Breeding density in the peregrine and raven. *Ibis* **104**, 13–39.

—— (1963). The status of the peregrine in Great Britain. *Bird Study* **10**, 56–90.

—— (1965*a*). Organo-chlorine residues in some raptor and corvid eggs from northern Britain. *Br. Birds* **58**, 65–81.

—— (1965*b*). The peregrine situation in Great Britain 1963–64. *Bird Study* **12**, 66–82.

—— (1967*a*). Decrease in eggshell weight in certain birds of prey. *Nature. Lond.* **215**, 208–10.

—— (1967*b*). The peregrine situation in Great Britain 1965–66. *Bird Study* **14**, 238–46.

—— (1969). Population trends of the peregrine falcon in Great Britain. In *Peregrine falcon populations; their biology and decline* (ed. J. J. Hickey), pp. 239–69. University Press, Wisconsin.

—— (1970). Changes attributable to pesticides in egg breakage frequency and eggshell thickness in some British birds. *J. appl. Ecol* **7**, 67–115.

—— (1972). The peregrine population of Great Britain in 1971. *Bird Study* **19**, 117–56.

—— (1980). *The peregrine falcon.* Poyser, Calton.

Research Study Group (1961). *Toxic chemicals in agriculture and food storage.* HMSO, London.

Risebrough, R. (1970). More letters in the wind. *Environment* **12**, 16–27.

Risebrough, R. W., Huggett, R. J., Griffin, J. J. and Goldberg, E. D. (1968). Pesticides: transatlantic movements in the northwest Trades. *Science, N. Y.* **159**, 1233–6.

—— Rieche, P., Herman, S. G., Peakall, D. B. and Kirven, M. N. (1968). Polychlorinated biphenyls in the global ecosystem. *Nature, Lond.* **220**, 1098–1102.

Robinson, J. (1969). Organochlorine insecticides and bird populations in Britain. In *Chemical fallout. Current research on persistent pesticides* (ed. M. W. Miller & G. G. Berg). pp. 113–69. Charles C. Thomas, Springfield Illinois.

—— Brown, V. K. H., Richardson, A. and Roberts, M. (1967). Residues of dieldrin (HEOD) in the tissues of experimentally poisoned birds. *Life Sci.* **6**, 1207–20.

Roburn, J. (1965). A simple concentration cell technique for determining small amounts of halide ions and its use in the determination of residues of organo-chlorine pesticides. *Analyst, Lond.* **90,** 467.

Royal Commission on Environmental Pollution (1971). *First report.* Cmnd 4585.

—— (1974). *Fourth report.* Cmnd 5780.

—— (1979). *Seventh report. Agriculture and pollution.* Cmnd 7644.

—— (1984). *Tenth report.* Cmnd 9149.

Rudd, R. L. (1964). *Pesticides and the living landscape.* Wisconsin University Press.

—— and Genelly, R. E. (1956). Pesticides: their use and toxicity in relation to wildlife. State of California Department of Fish and Game. *Game Bull.,* **7.**

Secretary of State for the Home Department (1949). *Health, welfare, and safety in non-industrial employment: report by a Committee of Enquiry.* Cmd 7664.

Secretary of State for Local Government and Regional Planning (1970). *The protection of the environment: the fight against pollution.* Cmnd 4373.

Select Committee on Estimates (1958). *The Nature Conservancy. 7th report.* Session 1957–58, Paper 255.

—— (1961). *The Ministry of Agriculture, Fisheries and Food. 6th report.* Session 1960–61.

—— (1962). *The Ministry of Agriculture, Fisheries and Food. 2nd Special Report,* Session 1962–63.

Select Committee on Science and Technology (1980). *Scientific aspects of forestry.* House of Lords Paper 381, Session 1979–80, Vol. 2, 135.

Sheail, J. (1976). *Nature in trust: the history of nature conservation in Britain.* Blackie, Glasgow.

—— (1984). Nature reserves, national parks and post-war reconstruction in Britain, *Environ. Cons.* **11,** 29–34.

Stanley, P. I. and Bunyan, P. J. (1979). Hazards to wintering geese and other wildlife from the use of dieldrin, chlorfenvinphos and carbophenothion as wheat seed treatments. *Proc. R. Soc. Lond.* (B) **205,** 31–45.

Strickland, A. H. (1966). Some estimates of insecticide and fungicide usage in agriculture and horticulture in England and Wales, 1960–64. *J. appl. Ecol.* 3 Suppl. 3–13.

Study Group on Education and Field Biology (1963). *Science out or doors.* Longman, London.

Tait, E. J. (1976). Factors affecting the production and usage of pesticides in the United Kingdom. Ph.D. thesis, Cambridge (unpublished).

Tait, J. (1981). The flow of pesticides: industrial and farming perspectives. In *Progress in resource management and environmental planning* (ed. T. O'Riordan and R. K. Turner), pp. 219–50. Wiley, London.

Tatton, J. O. G. and Ruzicka, J. H. A. (1967). Organochlorine pesticides in Antarctica. *Nature, Lond.* **215,** 346–8.

Taylor, J. C. and Blackmore, D. K. (1961). A short note on the heavy mortality in foxes during the winter 1959–60. *Vet. Rec.* **73,** 232–3.

Tinker, J. (1970). Pesticides—time for firm decisions. *New Scient.* **45,** 15–17.

Treleaven, R. B. (1961). Notes on the peregrine in Cornwall. *Br. Birds* **54,** 136–42.

Turtle, E. E., Taylor, A., Wright, E. N., Threarle, R. J. P., Egan, H., Evans, W. H. and Soutar, N. M. (1963). The effects on birds of certain chlorinated insecti-cides used as seed dressings. *J. Sci. Fd Agric.* **14,** 567–77.

Walker, C. H. (1966). Some chemical aspects of residue studies with DDT. *J. appl. Ecol.* 3 Suppl., 213–22.

—— and Mills, D. A. (1965). Organic chlorine insecticide residues in goosanders and red-breasted mergansers. *Wildfowl* **16**, 56–7.

—— Hamilton, G. A. and Harrison, R. B. (1967). Organochlorine insecticide residues in wild birds in Britain. *J. Sci. Fd Agric.* **18**, 123–9.

Way, J. M. (1965). Agricultural chemicals and wild plants. *New Scient.* **25**, 434–5.

—— (1968). Vegetation control and wildlife. *Proc. Br. Weed Control Conf.* 9th, 989–94.

—— (1969*a*). *Road verges: their function and management. Proceedings of symposium.* Nature Conservancy, Abbots Ripton.

—— (1969*b*). Toxicity and hazards to man, domestic animals, and wildlife from some commonly used auxin herbicides. *Residue Rev.* **26**, 37–62.

—— (1970*a*). Roads and the conservation of wildlife. *J. Instn. Highw. Engrs.* **17**, 5–11.

—— (1970*b*). Wildlife on the motorway. *New Scient.* **47**, 536–7.

—— (1973). *Road verges on rural roads: management and other factors.* Nature Conservancy, Abbots Ripton.

—— (1974). The management of grass. *Munic. Engng, Lond.* **151**, 1639–42.

—— (1976). *Grassed and planted areas by motorways.* Institute of Terrestrial Ecology, Abbots Ripton.

—— (1977). Roadside verges and conservation in Britain: a review. *Biol Cons.* **12**, 65–74.

——, Newman, J. F., Moore, N. W. and Knaggs, F. W. (1971). Some ecological effects of the use of paraquat for the control of weeds in small lakes. *J. appl. Ecol.* **8**, 509–32.

Wigglesworth, V. B. (1945). DDT and the balance of nature. *Atlant. Mon.* **176**, 107–13.

Wild Life Conservation Special Committee (1947). *Conservation of nature in England and Wales.* Cmd 7122.

Williams, R. (1977). Government response to man-made hazards. *Government & Opposition* **12**, 3–19.

Williams-Ellis, C. (1967). *Roads in the landscape.* HMSO, London.

Witt, J. B. de (1956). Toxicity of chlorinated insecticides to quail and pheasants. *Atlant. Nat.* 115–18.

Woodford, E. K. (1964). Weed control in arable crops. *Proc. Br. Weed Control Conf.* 7th, 944–62.

Working Party on Precautionary Measures against Toxic Chemicals used in Agriculture (1951). *Toxic chemicals in agriculture.* HMSO, London.

—— (1953). *Toxic chemicals in agriculture: residues in food.* HMSO, London.

—— (1955). *Toxic chemicals in agriculture: risks to wild life.* HMSO, London.

INDEX

Abbreviations

dieldrin (*contd*)
  passage from soil to birds 152, 155–6
  residues in soil, soil fauna 151, 152
  in seed-dressings 60–1, 63, 68, 73, 83
    accumulation, levels 73, 74
    see also seed-dressings
  in sheep-dips, eagle decline 116
  in small birds of prey 146, 147, 148
  withdrawal recommendation 106
Dieldrex 60–1, 63
dinitro-ortho-cresol (DNOC) 16, 54
  banning 24
  poisoning, human 22
  red spider mite predator killing 29
  trails on toxicity 26, 27
dioxin 125, 126
disposal,
  industrial effluent, PCBs 207, 210, 211
  pesticides, containers 180, 183
Diver, Captain Cyril 7, 8, 9, 24
DNOC, *see* dinitro-ortho-cresol
Doomsday syndrome, the 228
Douglas, Lord, of Barloch 183
drifting of spray 27, 48, 97
ducks 162–3
Duffey, E. 26, 32, 42, 46, 128
Duke of Edinburgh, Philip 225–6, 242

eagle,
  golden 115–16, 160–1, 162, 185
  white-tailed 207, 216
earthworms 151, 152, 156
ecosystem 48, 165
  effects of pesticides 165–6, 199
  passage of pesticide through 151–8
Edson, E.F. 119
educational role of Conservancy 175–6
Egginton, Joyce 232
eggs, toxic chemicals in 109, 116
  breaking, thinness, and sub-lethal levels
    160–1, 163–4, 167, 170
  distribution 161–2
  Further Review 173
  Gunn's evidence against 170
  shells, biochemical changes 162–3,
    170
  butterfly larvae 159
  golden eagles 115–16, 160, 186
  PCBs in 208
  peregrine falcon 109, 113, 146, 160
  of seabirds, as indicators 149
  sparrowhawk 145, 163
Encephalitis, in foxes 70, 71
England Committee, the 4, 43, 72, 104
entomology, economic 28–9
Environment 210
Environmental Pollution 208

environmental revolution 225–32, 242–3
  backlash against conservation 226, 230
  triggered by disasters 227
European Conservation Year 2, 179, 225,
    230, 238, 239
Expert Committee 140, 141, 179
exposure to accumulation relationship 155

Farm and Garden Chemicals Act (1967)
    186, 187, 198
farming,
  herbicides in 123–8
  post-war 18–19
Far, J. 50, 132, 181
fat reserves 153, 157
Field, the 200
field vole (*Microtus agrestis*) 204
Fifteenth International Ornithological
    Congress (1971) 149
fish 85, 141, 155
Fisher, James 214
Fisheries Laboratories 215
Fisons Ltd 98, 119
Fitzwilliam, Lord 68, 69
flea-beetle 91
fluoroacetamide 181–2
food chain 1, 155
  exposure to accumulation relationship
    155
  persistent pesticides in 90, 100, 118
Food and Drug Administration 97
food production 17, 18–19, 51, 91, 191
foodstuffs, residues in 22, 23, 56
forestry, herbicides in 123–8
Forestry Commission 98, 124, 152, 175
  Annual Report (1967–69) 124
  2,4,5-T use by 124, 125, 126
forestry ecosystem 152
fox-death 68–74, 77, 150
Frazer, A.C. 97
Frazer Research Committee, *see* Research
    Committee
Freshwater Fisheries Laboratory 209
frog (*Rana temporaria*) 159
fruit-trees 29
Fryer, J.D. 126, 127
fungicides 16, 58
*Further review of certain persistent
    organochlorine pesticides used in
    Great Britain* (1969) 168–74, 192,
    238

Game Research Station 60, 69, 77, 80, 98
gas-liquid chromatography 55, 99, 161,
    185, 206–7, 215, 222
geese, mortality 195–6

Gloucestershire roadside spray use 7, 8, 9
 proposal (1963) 14
 trials, participation in 10
golden eagle (*Aquila chryaetos*) 115–16,
 160–1, 162, 185
Gowers Committee 22
grebe,
 great-crested 147, 155
 western 86
Guardian, the 135, 214, 218
guillemot (*Uria aalge*) 203, 209, 212, 215,
 216, 221
Gunn, D.L. 170

habitats, trends affecting 48, 52
Hailsham, Lord 42, 44, 64
 Chemicals for the Gardener anomalies
  104
 Monks Wood and opening 40, 102
 on pesticide monitoring scheme 96
 in second Shackleton debate (1963) 93
 Silent spring, attitude to 91
 toxic chemicals unit 42, 44, 64
Haldane principle, the 4
*Handbook of the Society for the Promotion
 of Nature Reserves* 108
Haulm, potato 54
Health, Welfare and Safety in Non-
 Industrial Employment 22
hedgerows, study of 7–8, 25, 165
 requirement for experimental station 43
 see also roadside
hepatitis, acute 69
heptachlor 74, 83
 in badgers 150
 fox-death 74
 in peregrine falcon eggs 113
 in small birds of prey 148
 withdrawal recommendation 106
herbicides 19, 48, 123–39
 action, 16
 in aquatic systems 128–32
  code of conduct 129, 130
 in farming, forestry 123–8
 impact on wildlife 123
 most harmful 27
 motorway, roadsides 7, 8, 15, 135–9; *see
  also* roadside sprays,
 railway banks, cuttings 132–5
 value of 19, 48
heron (*Ardea cinerea*) 147, 149, 155, 162,
 208
Hickey, J.J. 141, 160
hill farmers 121
Himsworth, Sir Harold 90

Holdgate, M.W. 213, 214, 215, 219, 223,
 236
hormone changes 160, 162, 164
hormone weed killers 7, 9, 25
hover fly 10
human health, welfare 241–2
 persistent pesticides effect 90, 93, 172,
  242
 pesticides in sheep-dips 122
Huntingdon Working Party 40–6
Hunts 69, 70
Hurcomb, Lord 50, 95, 106, 133, 177

Ibis 109, 160
indicator species, choice of 143–51, 155,
 199
 badgers, bats 150–1
 birds 143–9
industrial pollutants 146, 172
 organo-fluorine 181–2
 wastes disposal 183, 184
Infestation Control Laboratory 33, 35, 39,
 46–7
 on dieldrex, advertisement 61
 on fox-death 69, 72, 73
 on organochlorine seed-dressings 61, 79,
  83, 192–3
 on organophosphorus seed-dressings
  196
 pigeon feeding trials 61, 73, 83
insect(s) 8, 10, 29
insecticides 16, 17
 benefits 19, 48, 91, 94, 239
 long-term effects 48, 49
 most harmful 27
 natural balance, effect 20
 resistance to 20
 review on effects (1949) 20–1
 synthetic contact 18
 universal 18
 zone of protection 58
 *see also individual insecticides, types of
  insecticides*
Insecticides, fungicides and the soil
 symposium 176
Institute of Terrestrial Ecology 236
integrated pest management (IPM) 166
Inter-department Advisory Committee on
 Poisonous Substances used in
 Agriculture and Food Storage, *see*
 Advisory Committee
International Committee for Bird
 Protection 103
international meetings 141, 142–3, 149
International Technological Conference on
 the Protection of Nature (1949) 20

persistent pesticides (*contd*)
  criteria of Notification Scheme 118–19
  Frazer Research Committee 97–101
  genetic, ecological problems 89
  President's Science Advisory Committee
    95–7
  second Shackleton debate 93–5
  Silent spring, response 87–92
  widespread, long term effects 49, 50,
    86–7, 116, 140, 168
  Wildlife Panel guide for (1963) 87
  see also organochlorine pesticides
Pest Control Ltd 18–19
pesticide 16
  annual introduction 231
  complacency over use 21, 49
  demand for 17–18, 19
  benefits, *see* benefits
  most harmful 27
  new, development times, costs 16–18,
    99, 230
  selective 16, 20, 21, 99
  side-effects 20, 21, 22
  wildlife debate 238, 239
  *see also individual pesticides and types of
    pesticides*
Pesticide Conspiracy, the 241
Pesticides Bill, proposed 188–9
Pesticides and the living landscape 140
Pesticides and pollution 177
Pesticides Safety Precautions Scheme 31,
    118, 140, 238
  carbophenothion seed-dressings 196
  criteria for, persistent pesticides 118–19
  non-agricultural uses, extension to 180,
    188
  organochlorine pesticides, ban 121, 191
    inadequacy 185, 192–4, 198
  publicity to 240
  replacement, statutory systems 188, 191,
    198
  Royal Commission report (1979) 197
  see also Notification Scheme
pesticide treadmill 165–6, 241
Pesticides and Wildlife 177
Philip, Duke of Edinburgh 225–6, 242
Pieris 165–6
pigeons, mortality 60–2, 64, 67, 83, 193
  feeding trials 61, 73, 83
  fox-death 68, 71, 73, 77
  homing, DDT on, 161
  sub-lethal pesticide effects 162
pine looper moth (*Bupalus piniaria*) 152
pipistrelles (*Pipistrellus pipistrellus*) 151
plant, species, susceptibility 11, 12
Plant Pathology Laboratory 97, 121
Plant Protection Ltd 98, 131

pollution 3, 16, 199–224, 230 242
  Environmental, Central Unit on 227
    heavy metal 204
    Irish Sea seabird wreck 199, 212–21
    monitoring of 221–4
    oil 202
    PCBs 146, 206–12, 215, 216, 218, 220
polychlorinated biphenyls (PCBs) 146,
    206–12, 227
  discovery as contaminant 206
  Irish Sea seabird wreck 215, 216, 218–21
  manufacturers, meeting Monks Wood
    210
  toxicity, lethal levels 209
  use of, widespread contamination 207–9
  voluntary withdrawals 210–12
population dynamics 159, 164
post-mortem analyses 52, 57, 61, 215
  birds, seed-dressing 61, 64, 67, 73, 99
  cats, foxes 71, 74
  Irish Sea seabird wreck 215, 216, 220
Potter, C. 30, 31
pre-clearance testing 196, 230
predatory species killing 20, 21, 29; *see
    also individual predatory species*
President's Science Advisory Committee
    (1963) 95–7, 221
Prestt, Ian 144, 149, 236
prey, behaviour of 154, 155, 159
Private Members Bill (1962) 185
Protection of Bird Act (1954) 28, 61, 110
Protection of the environment; the fight
    against pollution, White Paper, 1970
    2
public, education of 175, 176–8
pyrethrum 16

quail (*Colinus virginianus*) 112

railway banks, cuttings 132–5
Ratcliffe, D.A.,
  appointments 109, 111, 236
  egg-breaking, sub-lethal levels 160–1
  Fifteenth International Ornithological
    Congress 149
  golden eagle 115
  lethal levels, wild and caged birds 158
  peregrine falcon survey 109, 111–14,
    142, 146
  razorbills (*Alca torda*) 203
red spider mite 29
Rentokil Ltd 183
Research Committee (Frazer) 97–101
  annual report (1964) 98–9
  herbicide research, report 126, 127

274 INDEX